A SIMPLE GUIDE

ASTROPHYSICS SIMPLIFIED

MADHUR SOROUT

notionpress
.com

INDIA · SINGAPORE · MALAYSIA

Notion Press

No.8, 3rd Cross Street
CIT Colony, Mylapore
Chennai, Tamil Nadu – 600004

First Published by Notion Press 2020
Copyright © Madhur Sorout 2020
All Rights Reserved.

ISBN 978-1-64587-882-7

This book has been published with all efforts taken to make the material error-free after the consent of the author. However, the author and the publisher do not assume and hereby disclaim any liability to any party for any loss, damage, or disruption caused by errors or omissions, whether such errors or omissions result from negligence, accident, or any other cause.

While every effort has been made to avoid any mistake or omission, this publication is being sold on the condition and understanding that neither the author nor the publishers or printers would be liable in any manner to any person by reason of any mistake or omission in this publication or for any action taken or omitted to be taken or advice rendered or accepted on the basis of this work. For any defect in printing or binding the publishers will be liable only to replace the defective copy by another copy of this work then available.

Dedicated to the Memory of
Stephen Hawking

CONTENTS

AN INTRODUCTION

Some years ago, at the age of about 5, I was having a random discussion with my father (something about school, studies, friends etc.). I don't remember why or how but he asked me, "Have you ever thought why anything falls on the ground, towards the earth and does not go up in the sky?"

As a child, anyone would be very surprised, and so did I. I thought for a minute and then asked for the reason. Then he told me about Newton and his apple; that the Earth pulls everything towards it and this attraction is called '**Gravity**'. This was probably the simplest explanation of gravity for a five years old kid.

I asked him why the clouds are in the sky, and why the sky itself doesn't fall on the ground! He replied, "The clouds are very light and the sky is very far away". I was extremely "impressed" with gravity and Newton. It was only then when my interest in science was developed.

In the same year, he told me about some other random things in physics, including the Big Bang (when I asked him for the Big Ben and he misunderstood it with the Big Bang!).

After that, my interest in Physics and Astronomy kept on increasing. I am currently running my own science blog, '**Maddyz Physics**' (**maddyzphysics. com**).

Inspired by Stephen Hawking's 'A Brief History of Time' and Neil deGrasse Tyson's 'Astrophysics for People in a Hurry', I have written this book, 'Astrophysics Simplified: A Simple Guide to the Universe'.

This is a popular science (science for general audience) book. It mainly focuses on the major developments in science from Aristotle and Ptolemy (science was actually philosophy in the Aristotelian Era) to modern day Physicists like Stephen Hawking and Richard Muller. This is basically an introductory book on astrophysics for general readers.

Various concepts of Physics, like Relativity and Quantum Physics are explained in this book along with various topics in Astrophysics and Cosmology (mainly cosmology, which involves things like creation and evolution of the universe).

As the book is written mainly on Physics and related fields, it becomes very important to know what actually Physics is and what its importance is.

As we all know, humans have always been curious. This resulted in what we call 'Science' in the modern date. Physics is the oldest among all the sciences (in the form of Astronomy).

Physics helps us to understand the universe, nature (and natural phenomena) and the laws which govern this universe. Whether it is the world inside us, the world around us or the world beyond us, everything in this universe runs on some fundamental laws of Physics, which we understand through Mathematics. One can safely refer to Mathematics as the "Language of Physics" (or, even the language of the universe).

There was almost negligible application of mathematics on the natural phenomena until the seventeenth century. Galileo Galilei and Isaac Newton were the first to apply mathematics on Natural Phenomena, giving 'birth' to the modern ideas of motion (actually, fundamentally all physics is just motion of something).

The more "technical" or "advanced" Physics (which is trending among general audience) was developed in the twentieth century which included the contribution of a number of Physicists, namely Albert Einstein, Werner Heisenberg, Max Planck, Richard Feynman, Stephen Hawking, Roger Penrose and the list goes on.

Physics lays the foundation for modern Chemistry, Biology, Electronics, and Computer Science etc. If you 'boil anything down', you get Physics.

So, now as we have discussed what Physics is, we can discuss something about this book and its chapters. The book is broadly divided into three 'Phases' which are as follows:

1. **Phase One (Chapter 1):** This phase mainly consists of the development of science from Aristotle to Newton. This chapter is mainly about the basic meaning of 'motion' and 'force'.

2. **Phase Two (Chapter 2 to Chapter 7):** This phase consists of various concepts of Einstein's Theory of Relativity, both Special Relativity and General Relativity and some of their 'applications' and predictions. It also consists of various concepts of Quantum Mechanics.

3. **Phase Three (Chapter 8 to Chapter 9):** This phase consists of the creation of the universe and its evolution. This is the most important part of the book. This phase also consists of details about Dark Matter, Dark Energy, Time Travel, Black Holes and Wormholes. This phase mainly includes topics of Astrophysics and Cosmology.

You may observe that all the content in this book is not directly related to astrophysics, but everything in this book is used in astrophysics whether it is Newton's Laws, Relativity or Quantum Physics.

The book starts from various concepts of Classical Physics, the Theory of Relativity & Quantum Physics and then comes to the topics of Astrophysics and Cosmology, which, I think is the best way for a general reader to understand Astrophysics.

There is almost negligible mathematics in this book, but some major equations are included. I have done so because one cannot grasp the true 'beauty' in Physics without seeing the mathematical or abstract part alongside the practical laws.

[Physics is to mathematics what Tony Stark (Iron Man) is to J.A.R.V.I.S. or F.R.I.D.AY. The groundwork of computations, statistics, simulations etc. is done by F.R.I.D.A.Y. But, the real work of fighting the villains is done by Iron Man.]

Chapter one and three contain more mathematics than others but it is completely understandable by general readers as I have explained the meaning of every equation and **NEGLIGIBLE** mathematical symbols are used, the full names of the physical quantities are used instead, in most of the equations.

I have used the simplest language in the book and the simplest explanation of every concept that I could use. The reader only needs to know the basic mathematical operations (addition, subtraction, multiplication and division) and other basic concepts of algebra to understand the equations.

Another thing I want to clear is that I haven't explained every single topic of (basic) physics and astronomy. After reading this book, the reader is required to study the subject more to master it. There is much more to learn and explore about this subject. The book just gives the reader a feeling of how the subject 'looks' like. But, to help you in 'really learning the real physics', I have included a **"WHAT'S NEXT?"** section at the end of the book where you can find all the resources you need to learn Physics by yourself.

As I described, everything is explained in the "easiest way". **"A curious mind is all that is needed to use this book"**.

The aim of this book is to encourage the reader(s) to think in a scientific way and to ask 'big questions'.

Hope you all enjoy reading the book!

– Madhur Sorout

1
UNDERSTANDING NEWTONIAN PHYSICS

1.1 Introduction

Our current ideas of motion, rest and force were developed by Galileo Galilei and Isaac Newton in the seventeenth century laying the foundations of **'Classical Physics'** or **'Newtonian Physics'**. Newton had a very huge contribution in classical Physics including his 'Laws of Motion' and his 'Universal Law of Gravitation'.

On the other hand, 'Modern Physics' includes two major theories — 'Einstein's Theory of Relativity' (containing Special Relativity and General Relativity) and the 'Theory of Quantum Physics' (including quantum mechanics and quantum field theory).

Classical Physics is more complete in itself. But, the general theory of relativity and quantum field theory are quite incomplete or partial theories.

1.2 Motion, Rest and Force

Before discussing anything, we should know the basic definitions of motion, rest and force which are the basic concepts of Physics. One can't learn anything in Physics without knowing the 'real' meaning of motion and force.

Suppose there are two trains, 'A' and 'B', moving in the same direction with equal 'speeds', and, a person standing is stationary on the platform. Now, what can we say about the 'motion' of the train 'A'? Is it changing its position? For a person sitting in train 'B', train 'A' is not moving because both the trains are moving at the same speed. But, for the person standing on the platform, the train 'A' is in motion as it is changing its position with respect to him.

So, from the above example, we can conclude — an object is said to be in motion if it changes its position in some time interval with respect to any other object/observer. Motion is relative. But, what is time now? A clear definition of time can't be given. It is still something mysterious. We shall explore time in the forthcoming chapters. For this chapter, we can understand time as an 'interval' between two events. (Interval, again, is defined as the time period between two events, which doesn't seem to be a very helpful definition to us.)

It can be concluded that motion is always relative. But, now, how can we measure this motion? For measuring motion, we generally use two physical quantities, **'Speed'** and **'Velocity'**.

In simple words, speed is the measure of how much distance an object covers (distance is the 'actual' path travelled) in a given time period. Mathematically, speed is equal to the distance travelled divided by the time taken to cover that particular distance.

Speed = Distance/Time Interval

As we measure distance in **'meters' (m)** and time in **'seconds' (s)**, the unit in which speed is measured is **'meters per second'** or **'m/s'**.

Velocity is nothing but **'speed with direction'**. More precisely, velocity can be defined as the rate at which position changes with time. The term 'change in position' is often referred to as **'displacement'**. So, mathematically, velocity is equal to the change in position divided by the time interval in which the position changed.

Velocity = Change in Position/Time Interval

The units of both speed and velocity are the same, that is, **m/s.** Loosely speaking, if we define the direction of speed, it 'becomes' velocity. For example, if we say a man is running with **10 m/s,** it is his speed, but, if we say the man is running with **10 m/s in the north direction,** it is his velocity.

A difference in both quantities is that, as direction changes with time, the velocity changes, but the speed may or may not change. For example, **10 m/s in the north** and **10 m/s in the south** are two different values of velocities, but equal speeds.

Another notable difference between both the quantities is that speed is dependent on the total path covered by an object and velocity depends only on the initial and final positions of the object (*Final Position - Initial Position = Change in Position = Displacement*). For example, if an object travels in a circle (with a diameter of **7 meters**) in **2 seconds**, its speed will be **11 m/s**, but its velocity will be **zero** (as it returned to its initial position, and hence **displacement = 0**).

Now, the stuff we just discussed above is slightly wrong. Instead of using the terms 'speed' and 'velocity', we should have used the terms '**average speed**' and '**average velocity**' (for the above examples). The object travelled the circle of diameter **7 meters** (and hence, **circumference = 22 meters**) in **2 seconds**. It means that it travelled **22 meters** in **2 seconds**. And, to find out the speed, we just divided the distance by time, which gave us **11 m/s**, right? But, what if the 'speed' of the object was changing in these two seconds. Maybe, he covered **18 meters** in the first second and **4 meters** in the second second. Then, he will have two different speeds, **18 m/s** for the first half and **4 m/s** for the second half. But, the average speed for the total two seconds will be **11 m/s**. One other thing to notice is that even in a single second the speed may vary, so, we should say that his average speed in the first second is **18 m/s** and **4 m/s** in second second. The average speed, mathematically, is the total distance covered divided by the time taken to cover that (total) distance.

Average Speed = Total Distance/Time Interval

Similarly, the average velocity is the total change in position divided by the time taken to obtain that change.

Average Velocity = Total Change in Position/ Time Interval

Now, the thing is that how can we find the speed at a given 'instant' not in an 'interval'. For example, how can we calculate the speed of an object at **t = 4 seconds** not in the interval from **t = 0** to **t = 4 seconds**? When we have to find the speed of an object from **t = 0** to **t = 4 seconds**, we have the **time interval = 4 - 0 = 4 seconds**. So, we just divide the distance covered in those **4 seconds** by **4** and obtain the average speed of the object in the interval of 4 seconds. Now, to find the speed at **t = 4 seconds**, we have to find the '**instantaneous speed**' of the object at **t = 4 seconds**. To find the instantaneous speed, we have to divide the distance covered in a very small time interval by that time interval. The smaller the interval, the more accurate our answer. This is because of the fact that in a very small time interval, the chances of the speed to get changed is negligible. So, mathematically, instantaneous speed is equal to the small distance covered divided by the small time taken to cover that distance.

Instantaneous Speed = Small Distance/Small Time

Now, if we have to find the speed at **t = 4 seconds**, we shall observe the distance covered in a small interval before or after **4 seconds**. For example, if the object travels a meter in the next **0.001 seconds** (after **4 seconds** of its motion), its speed at **t = 4 seconds** will be **1/0.001 = 1000 m/s**. This is still not completely accurate (as the speed could have changed by some amount in this period of **0.001 seconds**), but we have an almost accurate answer.

Same is the case with instantaneous velocity.

Instantaneous Velocity = Small Change in Position/ Small Time

Scientifically speaking, when we use the terms 'speed' and 'velocity', it generally means 'instantaneous speed' and 'instantaneous velocity', not average speed and average velocity.

A thing that has to be clarified is that when an object is travelling with uniform speed (when it covers equal distances in equal time periods), its instantaneous and average speeds are the same for any instant or interval. Similarly, when an object has equal displacements in equal time periods, its instantaneous and average velocities are the same for any instant or interval.

Now we are in a position to discuss some other things about this concept. For example, how can we find the speed when we have a relation for distance and time? Suppose the distance travelled,'d' and time 't' for a particular motion are related as:

$$d = 5t^2$$

The above equation means that after a time 't' (from the beginning of motion), the distance (from the starting point), 'd' (which is in meters) is equal to '$5t^2$' ('t' is in seconds) for the motion in consideration (actually, this equation is the equation of motion for any falling object on Earth, falling from a height much smaller than the radius of Earth). For example, after **1 second**, the object travels a distance of **5 meters** (just replace 't' by '1'). Now, we have to find the speed of this object, more precisely, we have to obtain a relation between the speed at any time. From this equation, we can really find that the motion of this object is not a uniform motion [$(t = 1, d = 5)$; $(t = 2, d = 20)$; $(t = 3, d = 45)$], in the first interval of one second, the distance is **20 - 5 = 15 meters** and in the second interval of one second, the distance is **45 - 20 = 25 meters**. The motion is, hence, not uniform, so, we can't find its average speed. So, our task is to calculate the instantaneous speed of this object at any time. For this, we need a 'small distance' and 'small time'. So, suppose after 't' time, in a small time interval 'x', the object travels a small distance of 'y'. So, what we have to do is to replace 'd' by $(d + y)$ and 't' by $(t + x)$. The equation then becomes:

$$(d + y) = 5 (t + x)^2$$

What we have to calculate is y/x, which means the small distance 'y' divided by the small time period 'x'. And we shall use the above equation for calculating this. So, to solve this equation further, we have to have a basic knowledge of algebra. Simplifying this equation, we get:

$$(d + y) = 5 (t^2 + x^2 + 2xt)$$

Now, 'x' is a very small number, so its square is almost equal to zero. The square of a number lying between zero and one is smaller than the number itself. For example, the square of **0.01** is **0.0001**, which is almost equal to zero. Similarly, the square of 'x', 'x^2' is also very small, so it can be neglected. So, we get:

$$(d + y) = 5 (t^2 + 2xt)$$

Simplifying the equation, we get:

$$d + y = 5t^2 + 10xt$$

Now, we know that $d = 5t^2$, according to the equation of this particular motion. So, replacing 'd' by '$5t^2$', we get:

$$5t^2 + y = 5t^2 + 10xt$$

Subtracting '$5t^2$' from both sides of this equation (according to an axiom of Euclid, when equals are subtracted from equals, the results are equals), we get:

$$y = 10xt$$

Now, dividing both sides of the equation by 'x' will do our work:

$$y/x = 10t$$

And, the ratio 'y/x' is nothing but small distance divided by small time, which is speed. So:

$$Speed = 10t$$

This equation tells us that for this motion, the speed any instant, 't' is equal to *10 times* 't'. The speed in the above equation is in **meters per second**.

This is the basic method of finding speed or velocity when a relation between distance or position with time is given to us. But, there can be some cases when this relation is 'too complex' to be solved by this basic method, so, we've a whole branch of mathematics dealing with 'rates of change, small quantities etc.' known as 'Calculus'. Calculus was invented independently by Gottfried Leibniz and Isaac Newton. We don't currently have the 'space' here to discuss calculus further than this, so we may stop our discussion of calculus at this point!

Another concept which has to be discussed is 'finding the direction of velocity'. The above equation of motion describes the distance fallen after any time of any object falling from a small height (as compared to radius of earth) on Earth. In this case, the 'magnitude' (amount) of velocity will be the same as the speed (the magnitude of instantaneous velocity is equal to instantaneous speed for any kind of motion. And, for motion in a straight line, in a single direction, the magnitude of average velocity is also equal to average speed.). To find the direction of velocity, we have to find the displacement. The direction of displacement and velocity will be the same as velocity is displacement divided by time. In this case, the final position of the falling object is 'downwards' (relative to the initial position). The direction of displacement is specified by the straight line joining the initial position to the final position. In this case, the direction of displacement is vertically downwards. So, the direction of velocity in this case is also vertically downwards.

(Now, if we talk about the state of **'Rest'**, it is nothing but a special case of motion, with **'velocity equals zero'**, when there is no change in position of an object with respect to the observer.)

Another quantity for measuring motion is **'Acceleration'.** It is actually the rate at which the velocity changes with time. Mathematically, acceleration is the change in velocity (*final velocity - initial velocity*) divided by the time interval in which the change is obtained.

Acceleration = (Change in Velocity)/
(Time Interval)

If velocity remains constant, acceleration will be equal to zero as there will be no 'change' in velocity.

Acceleration also has a direction like displacement and velocity. The direction of acceleration is the same as the direction of 'change in velocity'. For example, if there is a positive change in (magnitude of) velocity (velocity increases), the directions of velocity and acceleration are the same, but if the (magnitude of) velocity decreases (change is negative), then the directions of velocity and acceleration are the opposite.

Like instantaneous and average speed, there is also a concept of instantaneous and average acceleration. In most of the cases, the acceleration is constant or uniform, so instantaneous and average velocities are the same in many cases. But, if it is non-uniform, instantaneous and average velocities are unequal. For calculating accurate instantaneous acceleration, we need a very small change in velocity and a very small time interval (similar to that of calculating instantaneous velocity).

So we can write:

Instantaneous Acceleration = (Small Change in Velocity)/(Small Time)

Scientifically speaking (again), when we use the term 'acceleration', we generally refer to instantaneous acceleration.

We just deduced that for a falling object, the relation between velocity and time is:

$$v = 10t$$

In this equation, 'v' (which is in meters per second) is the velocity (the direction, as we know, is downwards). What we have to do now is to calculate the acceleration for this object. Suppose, after a small time 'x' the velocity (positively) changes by 'z'. Replacing 't' by '$(t + x)$' and 'v' by '$(v + z)$' in the above equation, we get:

$$(v + z) = 10(t + x)$$

What we have to do now is to calculate 'z/x' which is the acceleration for this motion. Simplify the equation a little bit and we obtain:

$$v + z = 10t + 10x$$

As we know that, '$v = 10t$', we can replace 'v' by '$10t$' to get:

$$10t + z = 10t + 10x$$

Subtracting '$10t$' from both sides of this equation, we get:

$$z = 10x$$

Dividing both sides of the equation by 'x', we get:

$$z/x = 10 \ m/s^2$$

'z/x' is nothing but the acceleration (in meters per second squared, m/s^2, which is the standard unit of acceleration), so:

$$Acceleration, \ a = 10 \ m/s^2$$

One thing we can notice here is that we've got an 'absolute' or constant value of acceleration; it's not dependent on time, position, velocity etc. It is $10 \ m/s^2$ for this particular motion at any time or position. As the acceleration is constant, we can find the same answer by calculating average acceleration which will be the same for any time interval and for any instant. As the relation between velocity and time is:

$$v = 10t$$

To find the average acceleration, we have to put different values of 't', get a time interval and change in velocity, and then just divide the two. Firstly, put $t = 0$, we get $v = 0$. Put $t = 1$, and we get $v = 10$. The time interval is $1 - 0 = 1$, and change in velocity is $10 - 0 = 10$.

Dividing change in velocity by time interval, we get the same acceleration, $a = 10\ m/s^2$.

Now, **'Force'** is a push or pull which changes or tends to change the position (velocity, more precisely) of an object. However, there is a more accurate definition of force which is given by the second law of motion, which will be discussed soon. It is a misconception that a force causes or 'creates' motion. But, motion is already there. A force always changes the state of motion but it may or may not set an object in motion. It doesn't 'cause' motion. We should take an example to make this concept clearer.

For example, suppose a person standing stationary in space and no force is acting on it. An observer, moving with some velocity, passes by it. Now, with respect to that observer, the object is in motion, but no force acted on it.

It also doesn't cause motion in 'another' sense. The motion need not be in the direction of the force being applied. Suppose you throw a ball upwards (by applying an upward force), its motion is, obviously, in the upward direction. But, there is another force working on it, which is the gravity of earth. Now, the direction of the force of gravity is downwards, but the motion is upwards. Nevertheless, after sometime, the ball will stop and start to fall downwards, in the direction of force. But, now, suppose that you have enough 'power' to throw the ball (upwards) with a velocity of **11.2 km/s** which, luckily, is the escape velocity of earth (minimum velocity required to 'escape' the gravitational field of earth). At this velocity, the ball would just escape the earth. But, the gravitational force of earth is still being applied on the ball in the opposite direction. There was a force which set the ball in motion, but the other force, gravity, had 'nothing to do' with the object's state of motion or rest once the ball is far away from earth.

In a nutshell, a force is not responsible for 'causing' (uniform) motion, but is responsible for acceleration (this will be discussed very soon in this chapter).

1.3 Geocentric and Heliocentric Theories

Before the 'era' of Galileo and Newton, most people believed Ptolemy and Aristotle who said that the Earth is the centre of the universe because when one looks at the sky, the Sun, the Moon and stars seem to orbit the Earth (relative motion). (Even before that people believed that the earth is flat and rests on a tower of infinite turtles! However, there are still flat-earthers today. There is an 'official' Flat Earth Society also!).

At that time, telescopes weren't developed to see what was actually happening. So, Ptolemy and Aristotle thought that the Earth is the centre of the universe and every other object in the sky orbits the Earth.

There was another reason for thinking of Earth as the centre of the universe. When an object is dropped from a certain height, it is attracted towards the centre of the Earth. So, they thought that the centre of the Earth is the centre of the universe. This theory was called the **'Geocentric theory'** ('Geo' means Earth and 'centric' is used for centre).

Aristotle was the person who found that the Earth is not flat, but is (nearly) spherical. He presented this idea of a spherical earth in 330 BC.

Thirty years after this, in 300 BC, a Greek astronomer, Aristarchus, found that not everything is revolving around Earth, but, is revolving around the Sun also, that is, the Earth and the other five planets (at that time only five planets other than the Earth were discovered including Mars, Venus, Jupiter and Saturn), are revolving around the Sun, and, only the Moon is revolving around the Earth.

But, as Aristotle and Ptolemy were thought superior at that time, Aristarchus wasn't believed by most of the people.

It was not until the sixteenth century (around 1543), when a Polish astronomer and Mathematician, Nicolaus Copernicus, put forward the same theory as Aristarchus, which he called the **'Heliocentric Theory'**.

In his model, the Sun was placed at the centre of the universe and the planets orbited the Sun (and the Moon orbited the Earth). In the seventeenth century, an Italian astronomer, Galileo Galilei developed his telescope, and observed the moons of Jupiter, he found that the moons were orbiting Jupiter rather than Earth (as was described in the geocentric theory that everything orbits earth), so, his observations put an end to the geocentric theory and confirmed the heliocentric theory.

He actually not only confirmed heliocentric theory but also modified this theory. He excluded Sun as the centre of the universe and suggested that the pinpoints in the sky, which twinkled, were also like our Sun (this was also predicted by Aristarchus).

1.3.1 Kepler's Laws of Planetary Motion

At the same time Galileo proved the heliocentric theory (with modifications), a German Astronomer, Johannes Kepler presented his **'Laws of Planetary Motion'** which are as follows:

1. The planets orbit the Sun in elliptical (oval) orbits and the Sun is located at one of its two foci [the mid-point of semi-major axis (Fig 1.2) is called focus and there are two such foci].

2. A 'virtual' line segment joining a planet and the Sun sweeps out equal areas during equal intervals of time. (Fig 1.1)

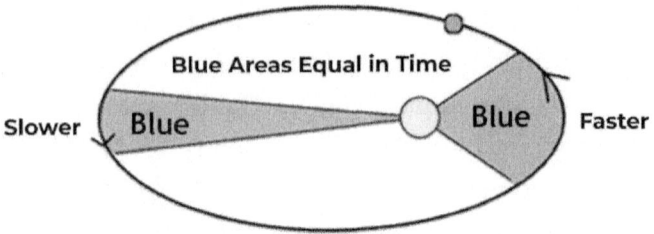

Fig 1.1 (above) - Illustration of Kepler's Second Law of Motion

3. The square of the time period of one revolution of a planet is directly proportional to the cube of the semi-major axis of its orbit (Fig 1.2). This means that the greater the length of the semi-major axis, the greater the time period of one revolution for a planet. The ratio of square of time period to the cube of semi-major axis is constant for all the planets in our solar system.

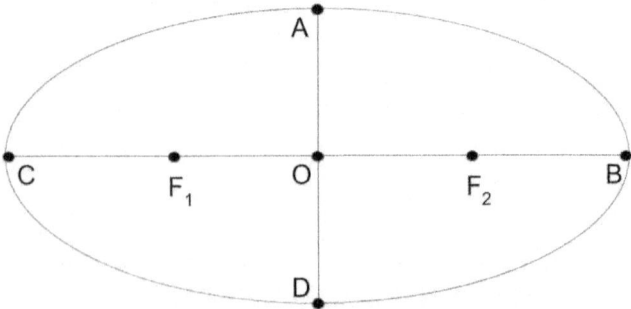

Fig 1.2 (above) - OB and OC are semi major axes of the oval (ellipse) above, OA and OD are semi minor axes of the ellipse and the mid- points of OC (F_1) and OB (F_2) are the two foci of the oval.

The laws of Kepler were later used and confirmed by Sir Isaac Newton.

1.4 Newton's Laws of Motion

Isaac Newton provided three **'Laws of Motion'** in which the second law is one of the most fundamental laws of Physics.

To understand Newton's Laws, we should firstly discuss the concept of **'Inertia'**.

1.4.1 Inertia and Galileo's Experiment

Galileo discovered a property that every object with a mass possesses, known as **'Inertia'**. He did a simple experiment (illustrated in Fig 1.3), which is as follows:

1. He placed two slopes opposite to each other.

2. The slopes were friction-less (friction is a force which opposes the motion of an object on a surface).

3. He placed a ball on one slope and the ball came down because of gravitation (or gravity).

4. Now, the ball went up on the second slope and it reached the same height from which it was left (or, say dropped).

5. Then he decreased the angle of the second slope and repeated this experiment, the result

was the same, the ball reached the same height from which it was dropped.

6. Then he reduced the angle of the second slope to zero. Now, the ball didn't stop and continued to move in order to achieve its original height.

He concluded from this observation that every object resists the change in its state of motion and this property of an object is called Inertia. Greater the mass of a body, the greater its inertia.

If Friction were absent, then...

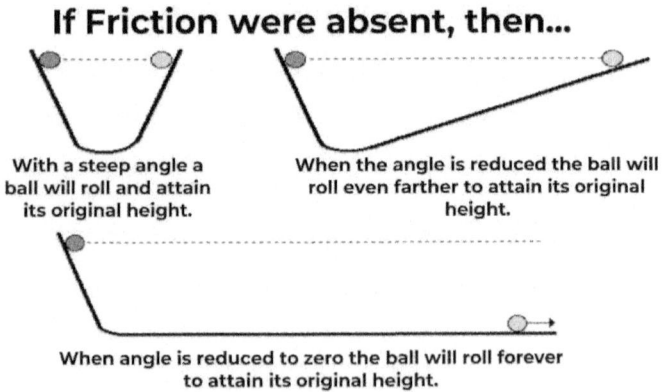

With a steep angle a ball will roll and attain its original height.

When the angle is reduced the ball will roll even farther to attain its original height.

When angle is reduced to zero the ball will roll forever to attain its original height.

Fig 1.3 (above) - Illustration of Galileo's experiment

There are also some daily life examples of inertia, like, when a car is at rest, it needs more force to accelerate it because it will resist the change in its state of rest. Another better example is the jerk felt by a person standing inside a bus when the bus suddenly stops moving. Actually, the legs of the person are connected directly to the floor of the bus. As the bus suddenly comes into rest, the legs of the person stops, and, as the upper part (which is not connected directly to the bus) was in motion, so it will resist the change in state of its motion and would like to remain in motion, and the person will feel a **'pseudo-force'** in forward direction due to inertia. This example is illustrated in Fig 1.4.

Another example of this kind of 'inertial pseudo-force' is the centrifugal force acting on a body in rotational/circular motion. It 'generates' due to the change in direction of motion (velocity) of the object. It is discussed in more detail in the last chapter, "RELATIVITY AND TIME TRAVEL".

So, there is always a pseudo force acting on the objects in an accelerating body, and this force is in the opposite direction of the acceleration of that body.

Riding a Bus

When a moving bus halts, you continue moving forward due to inertia.

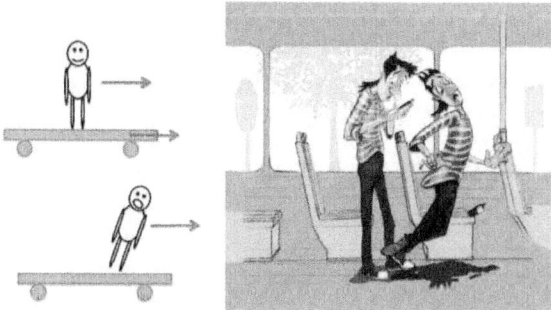

Fig 1.4 (above) - Example of Inertia

Aristotle believed that an object having large mass would fall faster than an object having a comparatively smaller mass. The reason behind this, that he believed, was that an object having a large mass would have a greater pull towards the Earth than an object of small mass.

An object having large mass does fall faster than an object having smaller mass (most of the time) when it is dropped from a certain height on the Earth because (most of the time) the object having smaller mass is resisted by air resistance (or air friction) and the object having large mass would experience smaller air friction. For example, if you take two lead balls of different masses and drop them from equal heights (at the same time), they will definitely hit the ground at the same time because the amount of air friction on them will be equal. Galileo also proved this fact (experimentally) that Earth accelerated everything at a constant rate.

We will see Newton's mathematical/theoretical reason for this very soon in this chapter.

1.4.2 Newton's First Law – The Law of Inertia

On the basis of Galileo's calculations, Newton formulated his first law of motion (also referred to as the law of inertia). It's more like a restatement of

Galileo's law, but still there is a little bit of difference between inertia and law of inertia.

This law states that a body will continue its state of motion or rest until a force acts on it. You might have seen some examples of this law in your daily life. When you make a ball roll on the ground and leave it, it continues its state of motion and stops only after some time because of the frictional force applied by the floor (Fig 1.5). If there were no friction, the ball would roll forever.

If you roll a ball on ground, it stops because frictional force acts on it. If there were no friction, the ball would roll forever if any force wouldn't be applied on it.

Fig 1.5 (above) - Illustration of Newton's first law of motion or law of inertia

1.4.3 Newton's Second Law – The Law of Action

To a rough approximation, a force can be defined as a push or pull which changes or tends to change the position or state of motion of an object.

Newton's Second Law of Motion provides a (mathematical) more precise definition of force.

It states that the force applied on an object is equal to the rate of change of its momentum.

'Momentum' is simply the product of an object's mass and its velocity. If we define momentum 'physically', momentum is the physical quantity which tells us that how much force an object will apply on another object (or experience a force by another object) when both the objects will collide with each other.

In simple words, it can also be defined as the **'combined effect or impact of mass and velocity'.** It can be understood by an example.

Suppose light is coming towards you from an electric bulb. Light's speed (or velocity) is the speed limit of the universe (we will discuss the reason behind it in next chapters) and light is very 'fast'. It can cover a distance of 300,000 kilometres in only a single second.

Light is composed of very small particles called **'Photons'.** Now, when the light will reach you, that is, when photons will collide with you, will you feel any force? The answer is certainly no. This is because photons have almost zero mass, so their momentum will also become almost zero and ultimately the force will also be negligible.

But now suppose a car of about 100 kilograms moving at a speed near (but smaller) to that of light, and it collides with another car at rest or moving towards the first car. This will, obviously, result in a huge damage.

This is because both the mass and velocity are very large, so momentum will be larger than both the quantities and force will also be huge.

Now, we should come to the term, **'Rate of change of momentum'.** It tells us how momentum changes with time. Just like velocity is the rate of change of position and acceleration is the rate of change of velocity.

It is nothing but the difference in the final momentum (momentum after the application of force) and the initial momentum (momentum before the application of force) and dividing this difference by the time interval in which the momentum changed. This gives us the value of the rate of change of momentum, and, the rate of change of momentum is equal to the force, according to the second law of motion.

So, mathematically, force is the change in momentum divided by the time interval to obtain that change.

Force = (Change in Momentum)/
(Time Interval)

Now, there is also the concept of 'instantaneous force'. It may be possible that the force being applied can change in the time interval we take, so, to calculate (almost) exact force at a particular instant, we need a small time interval.

So we can now write:

Instantaneous Force = (Small Change in Momentum)/ (Small Time)

Now, again, when the term 'force' is used, it generally refers to the 'instantaneous force'.

Now, we are going to prove that the force applied on an object is the product of its mass and the acceleration produced in the object due to that force.

We know that:

Force = (Change in Momentum)/ (Time Interval)

We know that momentum is mass times velocity. So, we can safely write:

Force = Change in (Mass ×Velocity)/ (Time Interval)

Now, we know that mass of an object is constant (at least in classical physics, when we are dealing with small velocities as compared to the velocity of light). So, what we are going to do now is to 'take mass out of the change'! So we can write:

Force = Mass × (Change in Velocity)/ (Time Interval)

Now an important thing to see here is that the change in velocity divided by time is nothing but the acceleration! So we can replace this with acceleration:

Force = Mass × Acceleration

So, the force applied on an object is simply equal to the mass of the object multiplied by the acceleration produced in it (due to the application of force on it). This is a very fundamental law of Physics. This above equation also concludes that a force always produces acceleration.

This equation really means that a force always produces an acceleration. Now acceleration doesn't only means speeding up, it also means slowing down (deceleration) and even change in direction of a moving object (because both velocity and acceleration are 'direction-dependent').

An important thing to discuss here is that this equation (of second law of motion) only tells us the relation of the acceleration of an object with the force applied, but doesn't really tell what is actually the force working here. For example, the force acting on the object may be electromagnetic or gravitational. We've different laws to calculate the force acting on an object in certain circumstances. Then we can equate that force with the mass of the object times its acceleration and do different things. In brief, the second law of motion is a universal law which is applicable for any type of force. If, for example, that force is gravity, then we can take the value of gravitational force as described by the law of gravitation and equate it with mass times acceleration to obtain new results.

We can also prove the first law of motion using the second law.

If the force applied on an object is zero (that is, no force is working on it), this means that whether the mass is zero or acceleration is zero. And, a force can only be applied on an object which has some mass, so, mass can't be zero. So, the acceleration will be zero, which means that the velocity hasn't changed, so, the object will continue its state of rest (velocity equals to zero), or uniform motion (that is, velocity will remain constant without any acceleration) if force is not applied on the object.

If same force is applied on two different objects of different masses then the lighter object will accelerate more than heavier and vice versa.

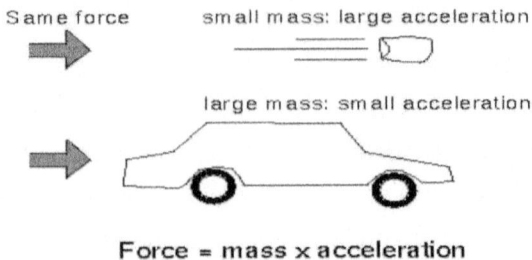

Same force   small mass: large acceleration

large mass: small acceleration

Force = mass × acceleration

Fig 1.6 (above) - Illustration of Newton's second law of motion or law of action

1.4.4 Newton's Third Law – The Law of Reaction

The third law of motion states that a force applied to an object has an opposite and equal reaction. In simple words, **"To every action, there is an equal and opposite reaction"**.

The action and reaction forces always act on two different objects. For example, when you walk you push the ground backwards (but it doesn't move due to having a very large mass), and, in turn, the ground pushes you forward. So, the action and reaction act on two different bodies and the directions of both are opposite. So it can be concluded that:

$$(Action\ Force) = - (Reaction\ Force)$$

And:

$$(Reaction\ Force) = - (Action\ Force)$$

There is a negative sign in the above equations to represent the opposite directions.

1.5 Newton's Theory of Gravity

Newton is best known for his **'Law of Universal Gravitation'**. This concept of gravity, started by Newton, unified the terrestrial with celestial. Newton said that the reason for an apple falling on Earth and the Moon orbiting the Earth is the same. This reason was **'Gravitation'** or simply **'Gravity'**.

Newton's theory of gravity states that every single object in the universe attracts every other object in the universe with a force which varies directly to the product of the masses of the two objects and this force varies inversely to the square of the distance between them.

This means that larger the masses of the two objects (or single one), the larger the force of gravity, and, larger the distance between the two objects, the smaller the force. For example, if the mass of any single body is doubled, the force between the two bodies will be doubled and if the distance between them is doubled, the force between them will reduce to one-fourth of the original force.

In simple words, it can be said that the force of gravity between any two objects **'A'** and **'B'** will be equal to the mass of **'B'** multiplied by the **'Intensity of the Gravitational Field'** of **'A'**.

Every object having mass possesses a gravitational field which is infinite in Newton's theory but the strength of the field goes on decreasing as the distance from the source of the gravitational field increases. So, it can be concluded, for any two objects **'A'** and **'B'**:

Force of Gravity between 'A' and 'B' = (Strength of the gravitational field of object 'A') × (Mass of object 'B')

In simple language, the above equation can be written as:

Force of Gravity = Mass × Strength of Gravitational Field

Also, according to Newton's second law of motion:

Force = Mass × Acceleration

In case of the force of gravity, the acceleration is **'acceleration due to gravity',** because any gravitational field accelerates any object uniformly (within small distances).

For example, earth accelerates everything (which falls from a certain height) at a specific acceleration of about **9.8 m/s^2.** It means that when an object falls from a certain height towards the earth, after every single second, its velocity increases by **9.8 m/s.** (**m/s** is the standard unit for measuring velocity and **m/s^2** is the standard unit for measuring acceleration.)

So, we can conclude:

Force of Gravity = Mass × Acceleration due to gravity

And, also:

Force of Gravity = Mass × Strength of Gravitational Field

So, to satisfy the above two relations, acceleration due to gravity must be equal to the strength of the gravitational field, that is:

Acceleration due to gravity = Strength of Gravitational Field

And this is the reason why every object on earth would fall at the same rate on earth, if there were no air friction or air resistance, because the acceleration due to gravity only depends upon the strength of gravitational field and not the mass of the falling object, in fact, the acceleration due to gravity is equal to the strength of the gravitational field. This means that the strength of the gravitational field is nothing but the measure of acceleration due to gravity. One thing to notice here is that the strength of the gravitational field, and hence, the acceleration due to gravity, decreases with increasing distance from the source of the field. So, for small heights from earth the value of acceleration due to gravity is **9.8 m/s^2**. But for larger heights, the value is somewhat smaller. But, still the value would remain the same for all the objects at a particular height irrespective of the mass of the object.

Actually, I want to show some stuff here. I want to derive the equation of motion of a falling object (on earth) which we used earlier in this chapter.

So, basically, one can easily prove that for any kind of straight line motion with constant acceleration, the general equation of motion is:

Distance Covered After time 't' = Half of Acceleration ×t^2

In this case the distance covered is 'height fallen' (let represent it by 'd') and the acceleration is **9.8 m/s^2** which we can, luckily, use as **10 m/s^2** (approximately). Half of **10** is, obviously, **5**. So, this equation then becomes:

$$d = 5t^2$$

Now, (coming again to the idea of gravitational field) the strength of the gravitational field of a celestial body (any body basically) is actually dependent on the mass of that body. So, we can say that the acceleration of a falling object (due to the gravity of any planet) depends upon the mass of the planet rather than that of the mass of the falling object or something.

The strength of gravitational field of an object is given by this equation:

Strength of Gravitational Field of an object = (Newton's Universal Gravitational Constant × Mass of the object)/(Distance from the object)^2

Newton's Universal Gravitational Constant is a Universal Constant (whose value is always constant) and its numerical value (value without unit) is **6.674 × 10^{-11}**.

All the above stuff we just discussed in this chapter is very simple, from a mathematical point of view. And the beauty of this is that using these simple equations of Newton's Laws, the law of gravity and basic equations of motion we discussed, one just can't only easily predict 'simple motions' like falling objects but also the whole orbit of a planet orbiting the sun (or any planet orbiting any star) to a very excellent approximation (but, we just can't do it here!).

2
THE MEANING OF RELATIVITY

2.1 Introduction

Classical Physics is only applicable to the objects which are far slower than light and are far larger than an atom.

This created a need for a new type of Physics which can explain any kind of motion of any 'kind of object' including very large scale structures like galaxies, black holes etc. and also the small-sized objects like electrons, quarks, photons etc.

This Physics was developed in the twentieth century (and still developing) and is called the 'Modern Physics' which consists of two major theories — **'Theory of Relativity'** or **'Relativity'** and **'Quantum Mechanics'**.

Relativity was developed by the famous German physicist, Albert Einstein. Some contributions of Hermann Minkowski, Hendrik Lorentz, Max Planck are also there.

Theory of Relativity comes in two different '*flavours*' — The '**Special Theory of Relativity**' and the '**General Theory of Relativity**' or simply special relativity and general relativity.

By name, special relativity seems to be more complex than general relativity, but actually, special relativity is much simpler than general relativity. Special relativity is called 'special' because it is applicable only on objects moving with a constant speed. This is a very rare or 'special' case because in our universe, hardly anything moves with a constant speed, almost everything is in acceleration in our universe.

General relativity is applicable to objects moving with a constant speed as well as accelerating objects, it's a 'generalization' of special relativity. Acceleration is a more common or a 'general' case in our universe.

Special relativity works (almost) well with particles but general relativity gives 'weird' results when applied on a very small scale (which will be discussed in upcoming chapters).

In this chapter, we shall discuss the basic principle of the special theory of relativity, its postulates and the '**Minkowski Space**' or '**Space-time**'.

2.2 What is Relativity?

Now, we should understand what actually the meaning of relativity is. Relativity means 'to be relative'.

It can be understood by the example given in the first chapter. Suppose you are inside a train and let the train is moving (on a straight track) with 40 mph (miles per hour) of speed. When you see a person sitting in the train in front of you, the speed of the person will seem to be zero for you but for an observer outside the train, the person will be moving with a speed of 40 mph.

So, with respect to you, the person is at rest, but, with respect to an observer outside the train, the person is in motion. This is the basic principle on which the theory of relativity is based on. There is no state of absolute motion or rest, everything is relative. So, relativity, to a **ROUGH** approximation, can be defined as –

"**Nothing is absolute in this universe, but everything is relative.**"

Only the speed of light is absolute, it doesn't depend on the speed or velocity of the observer, which was found experimentally by the "Michelson-Morley Experiment".

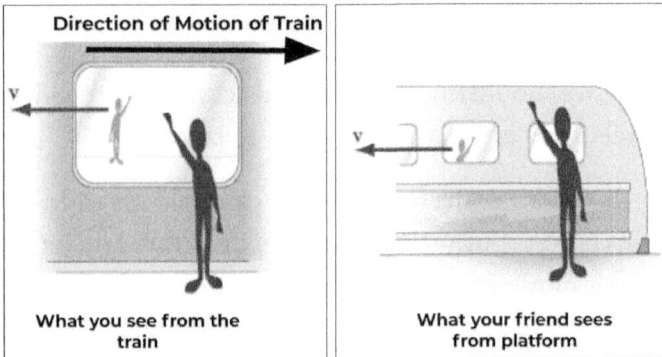

Fig. 2.1 (above) - Illustration of Relativity of Motion

2.3 The Special Theory of Relativity

As described above, all the special and general relativity is based on the basic idea that nothing is absolute in this universe, but everything is relative.

The special theory of relativity is mainly based on two postulates:

1. The fundamental laws of Physics don't change for any object (or observer) if it is at rest or moving with a constant speed. It means that rest and uniform motion are equivalent states (obviously, this was also known at the time of Galileo and Newton). This postulate is also known as the 'Principle of Relativity' which was also stated by Newton (but initially introduced by Galileo). Basically, you can't be sure if you are at rest or moving with a

constant speed. For example, if you are in a train moving in a straight line with constant speed (with respect to a person on ground) and all the windows are closed, there is no way for you to tell if the train is at rest or moving with some constant speed. If we have two observers, say, 'Abel' and 'Bruno', and, Abel is standing 'still' on the ground while Bruno is in a train moving with some velocity with respect to Abel. Now, for Abel, Bruno is moving, but there is an equal right for Bruno to say that he is at rest and instead Abel and everything else is just moving with that velocity in the opposite direction.

2. The speed (or velocity) of light remains constant for every observer and is independent of the velocity of the observer. This second postulate is just the result of the Michelson-Morley Experiment which (unintentionally) concluded that speed of light is a constant.

2.3.1 The Idea of Four-Dimensional Space-time

In this section, we shall discuss the '**Minkowski Space**' or simply '**Space-time**'. To understand what space-time is, we should really know what actually is '**Space**' and what is '**Time**'.

Space is a three-dimensional structure, that is, a structure made up of three dimensions and time is a single dimensional structure. Anyway, what is a '**Dimension**'?

Loosely speaking, if an object is '**n-dimensional**', then (minimum) '**n**' numbers are needed to specify the 'size' or extension of that object in space. For example, a 'line' is a **1-dimensional** object, so only one number, that is, its length is needed to specify its 'size'. If we have a cube or cuboid (which is a **3-dimensional** object), we need to find its length, breadth and height to specify its size. Then we have fancy things like a 'sphere' which is obviously 3-dimensional, but doesn't have length, breadth and height. So, what are the three numbers or 'quantities' which specify its size in space? So, it's basically just the diameter which specifies how 'long', 'wide' and 'deep' it is. You may have got a feeling of what a dimension is.

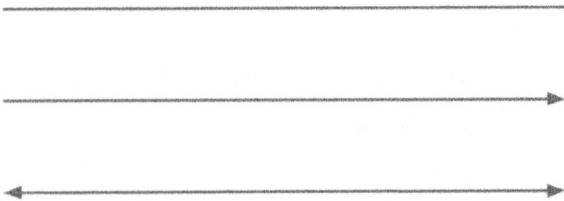

Fig 2.2 (above) {from up to down} - A line segment having a definite starting and ending point; a ray having definite starting point but not an ending point, it ends at infinity; a line having no definite starting or ending point, it starts from infinity and ends at infinity as well.

I would like to specify another thing here that in an '**n-dimensional**' space (space is just something where stuff happens), a minimum of '**n**' numbers are required to specify the position of an object and exactly '**n**' numbers are required to specify the position of a point.

Now, the dimensions of space are referred to as '**Spatial Dimensions**' (to differentiate it from the '**Time Dimension**'). According to relativity, space is **3-dimensional**, that is, there are three spatial dimensions in total. An object's position in space is measured in **metres (m)** and that in the time dimension is measured in **seconds (s)**. Basically, the position of an object in space is just the distance of that object from the given observer.

We can mark numbers on a line to define the position of an object (Fig 2.3):

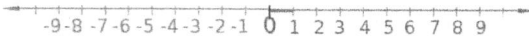

-9 -8 -7 -6 -5 -4 -3 -2 -1 0 1 2 3 4 5 6 7 8 9

Fig 2.3 (above) - These numbers are marked for defining position of any object

The point '**O**' is called '**Origin**' and is considered '**zero**', an object's position is measured with respect to the origin. In other words, the reference point we take to measure the position of an object is the origin. We can set it anywhere according to our requirements. Origin is just the point where the observer is sitting and 'seeing' things happen in space with respect to him.

Basically, a single dimensional space is represented in Fig 2.3. If we want to specify the position of a point in this kind of space, we will need only a single number, for example, if there is a point 'P' at the number '2', we can say the point 'P' is **2 units** away from the origin.

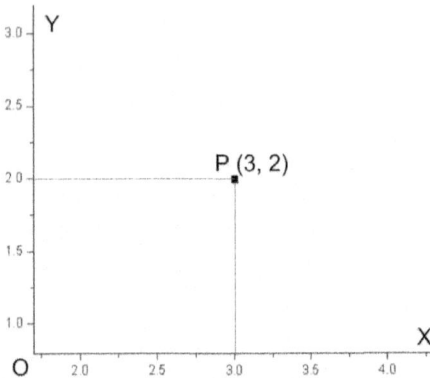

Fig 2.4 (left) - Illustration of a two-dimensional space

Similarly, we can represent a two-dimensional coordinate system, that is, a two-dimensional space, it can be done using two lines (or axes) [Fig 2.4].

In Fig 2.4, there are two lines (or '**axes**'), the '**X-axis**' and '**Y-axis**'. The **X-axis** is the horizontal axis and **Y-axis** is the vertical one. In one-dimensional space, there is only one axis either the horizontal **X-axis** or vertical **Y-axis** (or '**Z-axis**'). In Fig 2.4, we observe that the position of the point 'P' can be specified using two numbers, that is, **3** and **2**. The number (or '**coordinate**') of **X-axis** is written first, and then the coordinate of **Y-axis** is written, so, simply we can say that the position of point **P** is **(3, 2)**.

We can also represent a three-dimensional space like this but we'll need another horizontal axis, '**Z-axis**' for this (Fig 2.5).

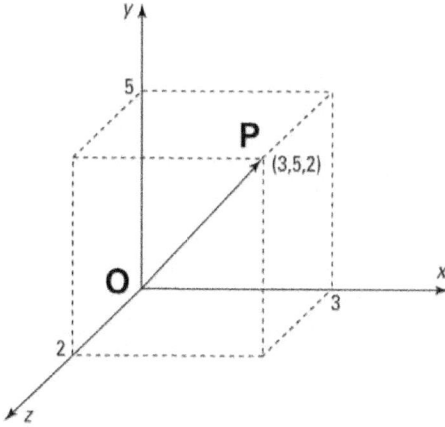

Fig 2.5 (above) - Illustration of a
three-dimensional space

In Fig 2.5, the position of the point '**P**' is specified using three numbers or coordinates, **3**, **5** and **2**, so it can be written as **(3, 5, 2)**. There are two horizontal axes and one vertical axis in the above image, all perpendicular to each other.

From the above observations, we can conclude what we stated: In an '**n-dimensional**' space, '**n**' numbers are needed to specify the position of a point, for example, in three-dimensional space, we need three numbers to specify the position of a point. Now, we may define 'dimensions' as the **number** of coordinates needed to specify the position of a point in a given space.

Now, we should briefly discuss '**absolute quantity**' and '**non-absolute quantity**'. An absolute quantity is a quantity which remains the same for all observers, (for example, the speed of light) and a non-absolute quantity can differ from observer to observer, depending on the observer's (or object's) velocity.

According to relativity, nothing is absolute, so, everything other than the speed of light (which was proved to be absolute both experimentally and mathematically/theoretically) should be considered as a non-absolute quantity.

Classical Physics ends the idea of absolute space (or distance), that is, the distance between two points can differ for two observers. A simple example of this is an object, like a ball placed at some random location. There are two observers, separated by a distance of one metre, interested in seeing the ball doing nothing. Maybe for one observer, the distance of the position of the ball is '**5 metres**'. Then for the second observer, it may be '**6**' or '**4 metres**'. This was a kind of no brainer thing. So, distance is not absolute.

We can now come on to the idea of '**Space-time**'. As we know, according to the second postulate of the special theory of relativity, velocity or speed of light is absolute, that is, every observer will measure the same speed of light. And, of course, *speed = distance/ time*, so, *speed of light = distance travelled by light/ time taken*, and, we know that the speed of light is absolute.

This situation can be possible in two cases, whether both distance and time are 'absolutes' or both are 'non-absolutes'. If they are non-absolutes, their values should be dependent on each other in such a way that the speed of light 'automatically' becomes absolute, and, because the distance is not absolute, so only the second situation is possible.

So, we can conclude that time measured by two observers can really differ. Every observer has its own measure of time now. Space (or distance) and time are now non-absolute quantities and both are dependent on each other.

This predicts another possibility which was put forward by Hermann Minkowski in 1908. Because space and time are dependent on each other, so both space (three-dimensional structure) and time (one-dimensional structure) are always fused with each other to form a single '**four-dimensional structure**' called '**Space-time**' or the '**Minkowski Space**' (Hermann Minkowski was one of the mathematics teachers of Albert Einstein).

There is a problem with a **4-dimensional** structure. We can't represent it as a coordinate system because we can't draw four lines perpendicular to each other. There is also one of the 'Euclid's postulates' that states **'only three lines can be drawn perpendicular to each other'**.

So, what we do is that we represent one or two spatial dimensions on one axis (either **X** or **Z**), and we represent time on the Y-axis. (We can also represent spatial dimensions on the other two axes and time on another axis). A 'geometrical' representation of four-dimensional space-time is illustrated in Fig 2.6.

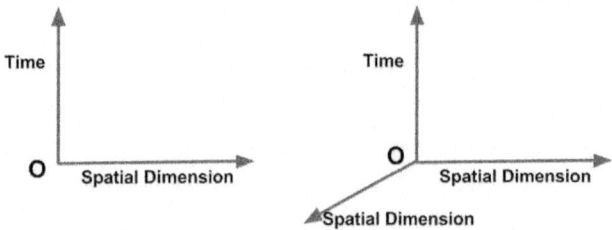

Fig 2.6 (above) - Illustration of 4-dimensional space-time as a coordinate system

In the next chapter, we shall discuss how the special theory of relativity predicts the '**contraction of space**' and '**dilation of time**', that is, how size decreases and time slows down for a moving observer and also we shall discuss the famous equation — $E = mc^2$.

3
THE RELATIVISTIC EFFECTS

3.1 Introduction

We now know that space (three-dimensional) and the dimension of time are always fused in a single four-dimensional structure called space-time or Minkowski space, and both space and time are non-absolute quantities, in fact, all quantities except the speed of light can be considered as 'non-absolutes'. It means that for an object at rest and an object in motion (relative to the object at rest), the measure of quantities like space, time and mass differ. In this chapter, we shall discuss the mathematical relations between these relative quantities.

(Note that these relations in special relativity are only applicable to objects at rest or moving with a constant velocity.)

3.2 Length Contraction – Squeezed Space

Consider a train moving with some (constant) velocity; a person is standing inside the train and another person is standing outside the train.

Suppose the person in the train has a torch and he switches it on when he is at the position 'A' (Fig 3.1). There is a distant screen 'O' in front of the person. Suppose light strikes the screen 'O' when the person in the train comes at the position 'B'.

Now, for the person outside the train, light would have covered a distance from 'A' to 'O', that is, 'AO' and, for the person inside the train, light would have covered a distance from 'B' to 'O' only, that is, 'BO'. This example is illustrated in Fig 3.1.

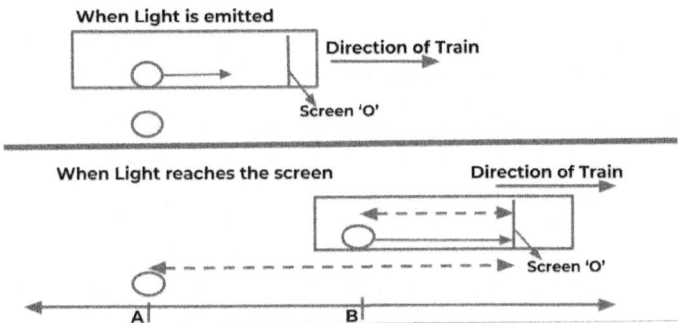

When Light is emitted

Direction of Train

Screen 'O'

When Light reaches the screen   Direction of Train

Screen 'O'

A   B

Fig 3.1 (above) - The dotted 'line' represents distance travelled by light with respect to both the observers and the 'ray' represents the light emitted.

From Fig 3.1, it is clear that the distance travelled by light is smaller for the moving observer as compared to the stationary observer. But, the difference in these distances will be almost zero at 'daily life' speeds or 'slow' speeds. 'Slow' here refers to far slower than the speed of light. Only at speeds near the speed of light, this difference in length can be observed.

This is not really length contraction, but is a demonstration of how distances and positions vary (in relativity) in two different frames (for different observers) with a relative motion between the two. Now, you may have got a feeling of how length contraction is going to 'look like'.

Now, we can just discuss the main concept. Consider two observers, 'Michael' and 'Freddie'. Freddie is moving with respect to Michael with some constant velocity (the opposite is also true, that is, Michael is moving with the equal velocity for Freddie, just in the opposite direction). Now suppose a rod is moving with the same velocity as Freddie (the rod is moving parallel to its length). What can we say about the length of the rod as measured by the two observers? As Freddie and the rod are moving with same speeds and in the same direction, there is no relative motion between the two. So, for Freddie the length of the rod is 'normal'. But, as there is some relative motion between Michael and the rod, Michael observes the length of the rod something less than Freddie does. Now, if the rod comes at rest with respect to Michael, the length of the rod will be 'normal' for Michael but less for Freddie, as there is some relative motion between Freddie and the rod now. One can just explain this 'without Freddie'. Suppose the rod is at rest for Michael at some time and the length of this rod for Michael is five meters. When the rod starts moving uniformly (with respect to Michael), the length (of rod) measured by Michael will be something less than five meters depending upon the velocity of the rod.

There is a proper mathematical relation between 'original length or size' and 'contracted length or size'. There is a factor called the *'Lorentz Factor'* (introduced by Hendrik Lorentz, a Physicist) which we see often in the equations of relativity. When the original size of the object (size at velocity equals zero) is divided by this factor, it gives the value of the contracted size (when it is moving). This means that:

Contracted Size = Original Size/Lorentz Factor

The value of the Lorentz Factor depends only on the (relative) velocity of the object being observed (this object, in our case, is the rod) with respect to the observer. The value of the Lorentz Factor is given by the following expression:

1/Square Root of [1 − (v²ᐟc²)]

It can be also written as:

$$\frac{1}{\sqrt{1 - \dfrac{v^2}{c^2}}}$$

In the above expression, *'v'* is the velocity of the object with respect to the observer and *'c'* is the speed of light and the value of the speed of light is **300000000 m/s**.

3.3 Time Dilation – Stretched Time

Time interval, like size, is also different for two different observers if they are moving 'for' each other (when I use the word 'for', I really mean 'with respect to'). Time runs slow for a moving observer. This can be understood with the same example given in Fig 3.1. For the moving observer, the distance travelled by light is shorter than that for the stationary observer. And, we know that:

Speed of Light = Distance Travelled by Light/ Time Taken

So:

Time Taken = Distance Travelled by Light/ Speed of Light

As the speed of light is constant for all observers, so, the time will differ. For the moving observer (in Fig 3.1), light took less time to reach the screen whereas for the stationary observer light took more time to reach the screen.

We can also say that less time has passed for the moving observer as compared to the stationary observer.

So, in simple words, time runs slowly for a moving observer as compared to a stationary observer. Now we should understand what this really means.

Consider again the two observers Michael and Freddie. Michael is standing still on the ground and Freddie is in a train moving with some constant speed 'for' Michael. Suppose Freddie is carrying a clock with him and Michael is keeping an eye on that clock. The difference between two 'ticks' of the clock is one second for Freddie. But, when Michael observes those two ticks, he finds that it is something more than one second. Now, if Michael is carrying a clock with him and Freddie is constantly observing Michael's clock; difference between two 'ticks' of the clock is one second for Michael. But, now, when Freddie observes it, he finds that the difference between two 'ticks' of the clock is more than one second! We now see a paradox here. In the first case, time was running slow for Freddie, but in the second case, time is running slow for Michael!

The answer is the following: Michael 'thinks' that time is running slow for Freddie, and, Freddie 'thinks' that time is running slow for Michael instead. This is because of the fact that both the guys are at rest for themselves. For Michael, Freddie is moving, and, for Freddie, Michael is moving. So, this is as simple as this.

This effect of '**Time Dilation**' has also been confirmed experimentally with '**Muons**' (an elementary particle). A muon is a type of particle which has a 'lifetime' of about **0.000002 seconds**. 'Lifetime of a Particle' means the time for which it remains in its 'original form' and after completing its lifetime, it decays into another type of particle.

Muons have a lifetime of about **0.000002 seconds** or **2 micro-seconds**, but it was found that muons travelling at ninety-eight per cent of the speed of light have their lifetime extended to five times, that is, a muon travelling at **98 %** of the speed of light lives five times longer than a stationary muon.

If we multiply the time interval between any two for a stationary observer by the Lorentz Factor, we obtain the value of the time interval between the same events 'for' a moving observer.

Time measured by a moving observer = Lorentz Factor × Time measured by a stationary observer

As described above the value of the Lorentz Factor depends only on the velocity of the observer.

But, before using this equation above, we have to keep in mind which observer is being considered a stationary observer and which one is being considered a moving observer.

Suppose an object is moving with some velocity. When the velocity increases, time becomes slower and slower for the object and when it reaches the speed of light, the time suddenly stops for the object, that is, time does not pass for the given object anymore.

However, anything having some 'rest mass' (mass of an object when its velocity is zero) can't travel at the speed of light or greater than the speed of light. The reason for this will be discussed later in this chapter.

Now, a question may arise, that if time doesn't pass for anything travelling at the speed of light, so does light itself experience time? The answer is no, light is made up of a type of particle called a '**Photon**', and a photon actually doesn't experience time (the rest mass of a photon is zero as well). This is the reason why it doesn't decay into other particles normally.

3.3.1 Twin Paradox

The most famous 'prediction' of time dilation is the '**Twin Paradox**'.

Suppose there is a pair of twins, '**A**' and '**B**'. **A** flies into outer space in a rocket with a speed nearly equal to the speed of light, and the other twin, **B**, remains at rest on Earth. When **A** returns to Earth, he finds that **B** is now older than him. The reason behind this is that when **A** will travel with almost the speed of light, time will start to run slower for him and thus, he will become younger than **B**.

But, now someone may argue that for **A**, **B** was the guy who was in motion, and thus, **B** should be younger than **A**, because the first postulate of relativity states that **A** can't really tell if he is the one who is moving if he travels at a constant speed with respect to **B,** and also **A** has the full right to say that **B** is moving. This is basically the paradox, who should be younger than whom?

This is the answer: **A** didn't really move at a constant speed 'all the time'. Initially, **A** and his rocket were at rest on Earth. Then, the rocket accelerated to some speed. When it accelerated, **A** really knew that he was moving as he would have felt a force downwards (similar to the force you feel when the car suddenly stops or accelerates suddenly). Also, to land back on Earth, the rocket would have been decelerated. So, it should be **A** who has to be younger.

3.4 Variation of Mass Due to Velocity

Mass of an object also varies with its velocity. The original mass of an object (at zero velocity) multiplied by the **Lorentz Factor**, gives us the increased mass, that is:

Increased Mass = Rest Mass × Lorentz Factor

Rest mass is nothing but the mass of the object when it is at rest.

But, according to the law of conservation of mass, mass can neither be created nor be destroyed, in other words, mass can't change. But in this case, actually, the total mass remains the same because the mass is not 'created', energy 'converts' into the mass.

According to special relativity, energy and mass are the same things and can be converted into each other. Mass is only a very dense form of energy. The mathematical relationship between mass and energy is given by Einstein's famous equation:

$$E = m \times c^2 \text{ or simply, } E = mc^2$$

In the above equation, 'E' is the energy, 'm' is mass and 'c' is the speed of light.

Any moving object possesses some '**Kinetic Energy**', and this energy increases with an increase in velocity. This energy converts into mass and hence, the mass increases. On a daily basis, this effect of increasing mass is not observable because velocity is far smaller than the speed of light. This effect is observable only when objects travel with velocities near the speed of light.

Now, we can discuss the reason for the speed of light being the 'speed limit of the universe'.

Suppose an object is travelling very close to the speed of light, because its velocity is very high, so its kinetic energy will also be very large, so now, much of the kinetic energy will be converted into mass, that is, most of the kinetic energy will be 'used up' to increase the mass, so, it will become difficult to increase the velocity further and it will never reach the speed of light.

It can also be understood in another way. We know that the value of the **Lorentz Factor** is:

$$1/Square\ Root\ of\ [1 - (v^2/c^2)]$$

When an object will travel at the speed of light, its velocity, 'v' will be equal to the speed of light, 'c' and hence, 'v^2' will be equal to 'c^2', and after solving the value of the **Lorentz Factor**, we will get the value of **Lorentz Factor** reaches infinity when 'v' is equal to 'c', and also:

Increased Mass = Original Mass × Lorentz Factor

Because the value of the **Lorentz Factor** is infinite, the mass of the object travelling at the speed of light will become infinite.

Also because:

Energy = Mass × c^2

So, energy will become infinite also.

This simply means that to accelerate an object to the speed of light, an infinite amount of energy is required, which (obviously) can't happen. This is the reason why nothing can travel with the speed of light.

3.4.1 Working of Atom Bomb

An atom bomb also works on Einstein's special theory of relativity. It works on the equation:

$$E = mc^2$$

This equation describes that energy and mass are interconvertible. When a large amount of energy converts into mass, a small amount of mass is formed and when even a small amount of mass is converted into energy, a huge amount of energy is formed. This is because of the fact that mass is a dense form of energy.

In an atomic bomb, a '**Neutron**' (a particle inside an atom) is 'fired' to an atom of '**Uranium**' (an element). By the impact of the neutron, the '**Nucleus**' of uranium (the centre of uranium atom) becomes unstable and divides into two nuclei of '**Caesium**' (an element) and '**Rubidium**' (another element) and also forms some other neutrons.

The sum of the mass of these two nuclei and the neutrons is slightly smaller than the mass of the original nucleus because some of the mass is converted into energy.

The neutrons released from original uranium are brought in contact with other uranium atoms forming more energy and this cycle is repeated many times. This creates a very huge amount of energy.

This energy is in the form of '**Gamma Rays**'. Gamma ray is the most energetic form of light (or '**Electromagnetic Radiation**', more precisely) and it can destroy or change the '**DNA**' (**Deoxyribonucleic Acid**) of a cell.

The working of a nuclear bomb (or an atom bomb) is illustrated in Fig 3.2.

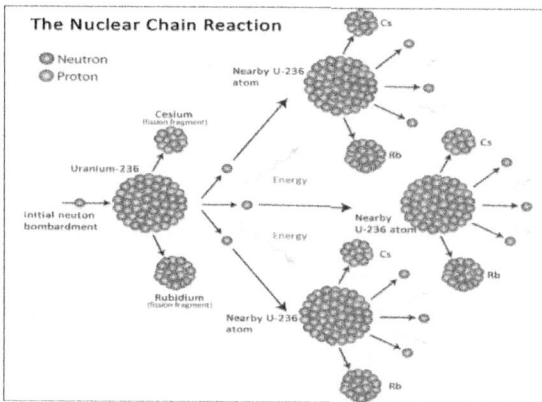

The Nuclear Chain Reaction

Neutron
Proton

Fig 3.2 (left) - Working of Nuclear Bomb

The next chapter is about 'Light Cones', another concept of the special theory of relativity.

4

LIGHT CONES

4.1 Introduction

An '**Event**', in relativity, is a definite point in space-time. An event is something which occurs at a definite position in space and at a definite time (having a particular place and time), and its position in space-time can be specified using four numbers (3 of space, 1 of time). An event has a definite future and a definite past as well and, in this chapter we are going to discuss things about this topic itself.

4.2 Relativity of Simultaneity

As the time interval between two events for two different observers can vary, this arises the possibility of 'simultaneity' being relative. Suppose for Michael, two events happened simultaneously. For Freddie, who is moving, the events will not be simultaneous. This may sound very 'scary'. But, here is an example to illustrate this.

Suppose Freddie is in a train which is moving with respect to the ground, and Michael is standing still on the ground. Freddie has two torches in his hand facing in the opposite directions (the torches are facing towards the two 'walls/ends' of the train). Freddie now switches the two torches on simultaneously (for him). Now, for Freddie, the two light beams emitted from the torches should hit the two 'walls' at the same time. But, for Michael, the train is moving. We also know that the speed of the two light beams will be the same for Michael (and Freddie as well) as the speed of light is constant for everyone. So, Michael sees this happening: one of the beams is 'running' towards the wall of the train and that wall is also 'running' towards the beam. The other beam is going towards the other wall, but that wall is 'running' away from the beam. So, for Michael, one of the beams hit the wall first and the other one hits the (other) wall after some time. If you didn't get anything about this, just see the figure below (**Fig 4.0**).

One end/wall of the train is moving towards ray 'A' and the other wall is moving away from the ray 'B'.

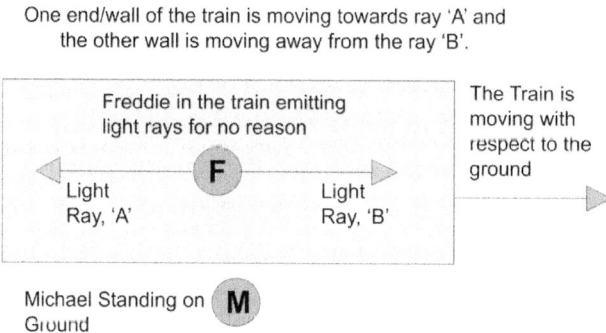

Freddie in the train emitting light rays for no reason

F

Light Ray, 'A'

Light Ray, 'B'

The Train is moving with respect to the ground

Michael Standing on Ground

M

Fig 4.0

Simple thing is that Michael sees ray 'A' covering less distance (so taking less time) and the ray 'B' covering more distance (so taking more time).

This whole idea raises the possibility of something more scary. Suppose for Michael, firstly an event 'A' takes place and, after some time another event 'B' takes place, then, if Freddie is moving fast enough, he could see 'B' happening before 'A'!

This can happen, but only in some conditions. Suppose 'A' is the event of an explosion of some bomb and 'B' is the event of a person being killed due to that explosion (relativity often uses violent examples). Now, if Freddie moves fast enough, he could see a person being killed without any reason (**B** takes place before **A**)! This creates a paradox. Luckily, we've figured out this stuff.

If the two events 'A' and 'B' are connected in such a way that 'B' is affected by 'A', then there is no possible speed with which Freddie can move to see 'B' happening before 'A'. This would be possible only if Freddie could move faster than light, which is not allowed. Freddie can only see 'B' happening before 'A' only if both these events are independent of each other. All this can be proved quite easily using elementary mathematics, but I won't do it here, as we don't have much space to do all that here! (if you really want to know the math which I skip in this book, you will find some awesome resources in the last part of this book, **"WHAT'S NEXT?"**)

All the things which are now going to be discussed in this chapter are related to this concept.

4.3 Light Cones

Suppose an event takes place in which light is emitted, and because light is a wave, it creates 'ripples' in 'electromagnetic field' as water waves create ripples in water (every 60 seconds in Africa, a minute passes!) (Fig 4.1).

In Fig 4.1, a wave is illustrated, whose source is the point '**A**', that is, it is generated from the point '**A**'. But, when this wave is represented with time, a cone is obtained (Fig 4.2).

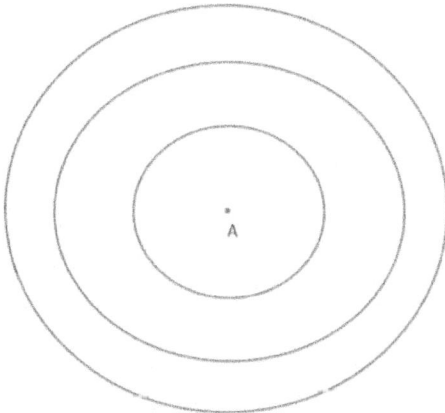

Fig 4.1 (above) - A wave. Its source is the point 'A'.

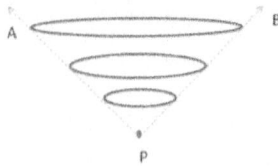

Fig. 4.2 (left) - A wave represented with time and two spatial dimensions.

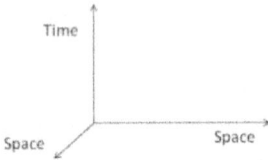

In Fig 4.2, a wave is represented with the time dimension and two spatial dimensions (dimensions of space). Time is represented on the **Y-axis** (vertical axis) and the spatial dimensions are represented on **X** and **Z** axes (horizontal axes).

Similarly, when an event occurs and light is emitted, we can construct two cones with the event as the apex of the two cones (Fig 4.3).

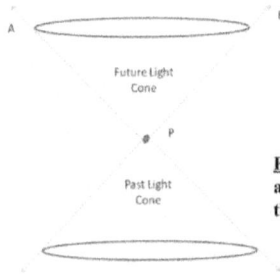

Fig. 4.3 (left) - The future and past light cones of the event 'P'.

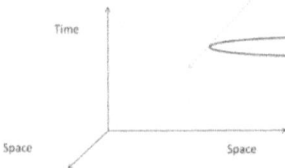

In Fig 4.3, the future light cone and past light cone of an event '**P**' are represented.

Now, we should really understand what this all really means. I just gave a very simple example to construct these light cones. This way is not really completely correct to do it. You can easily do it with the equations, which is the correct method.

But, nevertheless, we can really discuss what actually light cones are. The graphs in these light cone 'diagrams' are not plotted between space and time, but, instead, the **space** AND **speed of light multiplied by time**.

So, the boundary of these cones ('lateral surfaces of these cones') really represent the paths of light reaching '**P**' from the past and emitted from '**P**' to the future in just the 'right time'. This means the following: if any point lies on the boundary of, say the future light cone of '**P**', means that if an event is going to occur at the boundary of the future light cone (of '**P**'), and an observer sitting at '**P**' wants to stop that event from occurring, it has to travel with a minimum speed, which is exactly equal to the speed of light, to reach there in the right amount of time and do something to affect it; if that event lies inside the future light cone, the observer at '**P**' needs to travel with some minimum speed, which is smaller than the speed of light, to reach there in the right amount of time (a thing to be clarified here is that these cones are not real things which you can see, these are just graphs, just 'distance versus time' graphs with extra steps!).

Similarly, if an event occurs somewhere at the boundary (lateral surface) of the past light cone of 'P', and an observer wants to stop the event 'P' itself from occurring, it has to travel with a minimum speed, which is exactly equal to the speed of light, to reach 'P' in the right time and do something. And if that event lies inside the past light cone of 'P', the observer sitting at that event will require some minimum speed, which is something smaller than the speed of light, to reach 'P' in the right time.

If any event lies outside the light cones of 'P', the observer sitting at those events can't affect 'P' in any way because that would require them to travel at a speed greater than light. Similarly, any person sitting at 'P' can also not affect those events as it would require him/her to travel faster than light itself to reach those events.

In a nutshell, only the events in the past light cone of 'P' can affect 'P'. And, 'P' can affect only those events which lie in its future light cone. That is why the future light cone of anything is known as its **absolute future** and its past light cone is its **absolute past**.

There is an amazing example of this all. Suppose, due to any reason, our Sun suddenly disappears. If this happens, we shall not come to know about the Sun's disappearance suddenly because light travels at a finite speed. It will take about eight minutes to know about the Sun's disappearance because light takes eight minutes to travel from Sun to Earth.

This means that Earth will enter the future light cone of 'Sun's disappearance' after eight minutes of the disappearance of Sun (**Fig 4.4**). So, we will know about the darkness only after eight minutes of the Sun's disappearance.

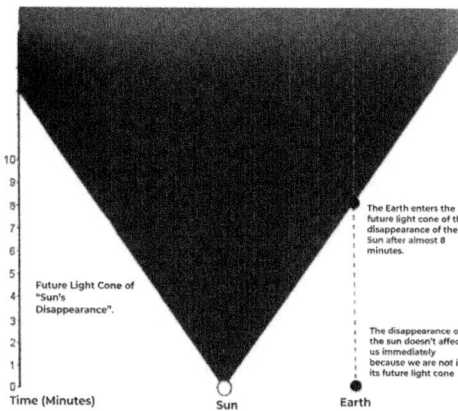

Fig 4.4 (left) - Disappearance of Sun

Special relativity describes this event of the Sun's disappearance in the terms of light waves, but the general theory of relativity can ALSO describe it in the terms of '**Gravitational Waves**' which will be discussed in chapter six.

Einstein knew that the special theory of relativity is only applicable to stationary objects and objects moving at a constant speed. He knew that his theory has to come for everything in the universe, that is, accelerating objects also. He needed a '**General Theory of Relativity**', which he finally developed in the year of 1915.

The general theory of relativity is the greatest achievement of the human mind till now. It is the greatest theory ever developed by a single person. It will be discussed in more detail in the next two chapters.

5

THE FOUNDATION OF
GENERAL RELATIVITY

5.1 Introduction

In the last section of the first chapter, Newton's theory of gravity was discussed which states that every single object in the universe attracts every other object with a force which varies directly to the product of the masses of the two objects and inversely to the square of the distance between them.

And the equation below is used to calculate this force.

Force of Gravity = Mass × Strength of Gravitational Field

We concluded in the first chapter:

Acceleration due to gravity = Strength of Gravitational Field

Now, Newton's theory gives wrong results for stronger gravitational fields (to make the strength of a gravitational field of a body 'stronger', one can either increase the mass of that body or bring any object 'closer' to that body. The main thing is that the strength of the gravitational field of any body is not the same everywhere.

Einstein then gave his model of gravity in his general theory of relativity which is still the most correct description of gravity and gives correct results for (almost) every situation. It has been found EXACTLY correct all the time in real observations.

In this chapter, the limitations of Newton's theory and the '**Equivalence Principle**', which is the basic principle of the general theory of relativity will be discussed.

5.2 Limitations of Newton's Theory of Gravity

This equation for the strength (or force) of gravity (which Newton has given us) is so accurate (for weak fields), that we still use it for practical purposes like flying a rocket from Earth and landing something on the moon or any other planet. But, there are two problems with Newton's theory – It fails for strong gravitational fields like a neutron star and the way Newton described gravity was not correct, in fact, Newton himself wasn't satisfied with his theory. He didn't know how gravity actually works. So, he '*invented*' something called '**Gravitational Pull**'. But, he didn't actually explain the reason for this gravitational pull. He didn't explain how gravity works basically.

We still use this equation of Newton because it is much simpler than the equations of general relativity. As Sun's gravitational field beyond planet Mercury is 'weak', we can calculate gravity between any planet (except Mercury) and Sun correctly using Newton's equation.

The orbits of all planets (except Mercury), predicted by general relativity are the same as those predicted by Newton's theory of gravity.

5.3 The Equivalence Principle

The general theory of relativity is the currently accepted model of gravity which describes gravity in a different way.

You may have searched for general relativity on the web and found this type of description – **"Gravity is a consequence of distortion of space-time"**.

But what Einstein meant by distortion of space-time and how he developed this idea?

Einstein used his equivalence principle to develop this idea. The principle of equivalence is the foundation of the general theory of relativity.

The equivalence principle states that — the effects in a gravitational field are the same as the effects in a uniformly accelerated body (analogous to the first postulate of special relativity). In other words, one can't be a hundred percent sure whether one is inside a (uniform) gravitational field or is being accelerated uniformly. A uniform gravitational field is a field whose strength is the same at all points, that is, the acceleration due to gravity is the same at any distance from the source of the gravitational field. A perfectly uniform gravitational field doesn't exist at all. The earth's gravitational field, for example, is almost uniform at very small heights (small as compared to the radius of earth). At same heights, the acceleration due to gravity is, obviously, the same for any object.

We can understand the idea of the equivalence principle with the help of the following examples:

1. A man is in a rocket at rest on Earth and he drops a ball in the rocket.

2. The man is now in the same rocket without any contact with a gravitational field and the rocket is accelerating at an acceleration of **9.8 m/s^2** (which is, luckily, the acceleration due to gravity on the surface of earth and at small heights.) Now, again he drops the ball.

He will see the ball falling in the same way in both the cases (Fig 5.1). So, we can conclude that the effects in the accelerated rocket are the same as the effects in a gravitational field. This is actually because of inertia. We discussed in the first chapter that a 'pseudo-force' is experienced by a person in a bus, when the bus accelerates.

The same concept is here. The rocket is accelerating upwards, so the person in the rocket will experience a pseudo-force in downwards direction.

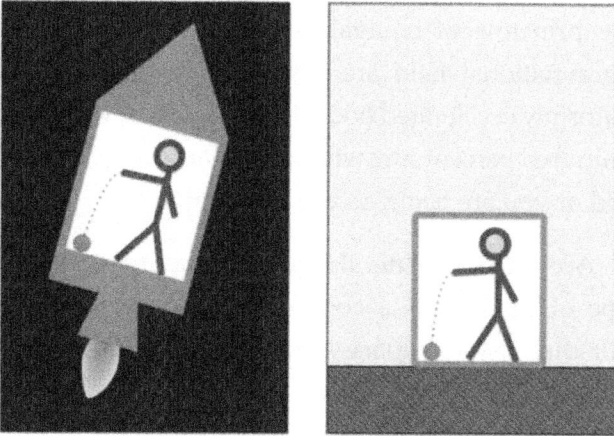

Fig 5.1 (above) - The equivalence principle.

This is what we call the 'Principle of Equivalence'. Einstein developed the most revolutionary idea of 'Curved Space-time' with this principle itself, which will be discussed in the next chapter.

6

CURVED SPACE-TIME AND GRAVITATIONAL WAVES

6.1 Introduction

The principle of equivalence states — the effects in a gravitational field are the same as the effects in a uniformly accelerated body. In other words, you can't be a hundred percent sure whether you are in a gravitational field or you are being accelerated uniformly.

According to general relativity, gravity is a different type of force; it is a consequence of 'curves' in the four-dimensional space-time. Before the introduction of the 1915 general relativity, space-time was thought to be independent of what was happening in it. But general relativity predicts that the presence of mass and energy in space-time does affect it.

6.2 Curved Space-Time and Geodesics

Einstein used the equivalence principle to predict that space-time curves. We can take an example to know how he did it.

Consider a rocket (which doesn't have any contact with a gravitational field) which is (at least) **300,000 kilometres** long, this means that light takes one second to reach from top to bottom (and vice versa) of the rocket and according to special relativity, light's speed will remain constant for everyone no matter what the speed of the rocket is.

Consider, currently it is at rest or moving with a constant speed. Now consider two people, '**A**' and '**B**' standing at the top and the bottom of the rocket respectively.

'**A**' sends two pulses or waves (or rays) of light to the person '**B**', at the bottom of the rocket at a time gap of **one second**, that is, he sends the second pulse of light after **one second** of sending the first pulse of light. The person '**B**' with no doubt will receive these two pulses at the time gap of **one second**.

Now, suppose that the rocket starts to accelerate and '**A**' gets out of the rocket, and is just above the rocket. Consider that '**A**' is at rest. Now, if '**A**' sends two pulses of light at a time gap of **one second**, the person, '**B**' standing at the bottom will receive the second pulse in less than **one second**.

Why does it happen? The answer is because the light has to travel less distance in this case as '**B**' is moving towards the light rays.

This can be understood by Fig 6.1.

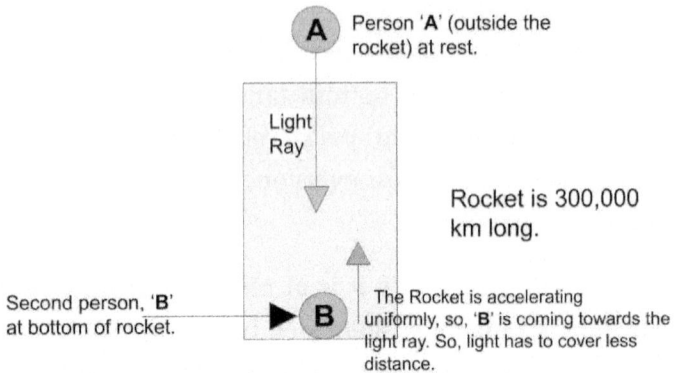

Person 'A' (outside the rocket) at rest.

Light Ray

Rocket is 300,000 km long.

Second person, 'B' at bottom of rocket.

The Rocket is accelerating uniformly, so, 'B' is coming towards the light ray. So, light has to cover less distance.

Fig 6.1 (above)

In Fig 6.1, the **arrow pointing downwards** represents the light ray coming towards the person **'B'**. The **arrow pointing upwards** represents the direction in which the rocket is moving/accelerating (with respect to a stationary observer).

The person **'B'** is moving towards the light ray and the light ray is moving towards the person, and, because the speed of light is absolute, the distance which the light ray has to travel is less than the length of the rocket. So, the person at the bottom will receive the two pulses in a time interval less than one second, that is, time is lengthened (or dilated or slowed down) for **'B'** because less time has passed for him.

This means that space has contracted (or shortened) for the person **'B'** at the bottom of the rocket and time has dilated (or lengthened) for him. But, what can we conclude from this?

According to the equivalence principle, the effects in accelerating bodies are the same as the effects in gravitational fields. So, we can conclude that in a gravitational field, space contracts and time dilates, in other words, near a source of mass or energy, space is contracted and time is dilated and this effect goes on decreasing as we move away from the source of mass (or energy).

'**Gravitational Time Dilation**' has been experimentally confirmed also using atomic clocks on airplanes. The clocks aboard the airplanes were slightly faster than clocks on the ground. This effect is so significant that the GPS satellites need to have their clocks corrected.

It should be noted that time dilates (and space contracts) for both uniformly moving objects (for an observer at rest) and accelerating objects .

But, how does this concept explain gravity? Now, consider the case of an apple falling on Earth. Why does it fall? The answer is because the earth has mass, so, near the earth, space is contracted and time is dilated, and, because space is contracted, it pushes the apple to the ground! This is how gravity works. This '**Gravitational Space Contraction**' and '**Gravitational Time Dilation**' is known as '**Curving of Space-time**'.

Space pushes objects only in some definite paths, the straight path between any two points in four-dimensional space-time. This straight path between two points is called 'Geodesics'.

All the objects in space-time follow straight lines in (four-dimensional) space-time or they follow the shortest path between any two points (because space pushes them to move in these straight paths) in space-time. This path is called geodesics.

Now, the question arises, if space pushes an object to travel in straight paths, why does the Earth and other planets are orbiting the Sun in elliptical orbits? We can take an example to understand this.

The surface of Earth (or anything) is two-dimensional and is curved because it is a sphere. Now, on a globe of Earth, if you mark two points far enough and join them, do you get a straight line after joining them? The answer is no. The line or shortest path between these two points is curved. Now, if a rocket flies over Earth in a straight path, its shadow on Earth, that is, its path on a two-dimensional surface will be curved. This means that the rocket is moving straight in three-dimensional space but its path seems to be curved on the two-dimensional surface of Earth.

This can be understood by Fig 6.2.

Fig 6.2 (above) - The rocket is moving straight in three-dimensional space but its path seems to be curved on two-dimensional surface of Earth.

This effect is the same for four-dimensional space-time and three-dimensional space. For example, space is pushing Earth and other planets to follow the geodesics, that is, the straight path in four-dimensional space-time but they seem to move in curved paths in three-dimensional space.

But when an apple falls, it moves in a straight path in three-dimensional space also, that is, the apple is not moving in a curved path in three-dimensional space when it falls. Why? When you will mark two very nearby points on a very large globe of Earth, and join them, you will find that this line or shortest path is straight. This means that the distance between the two points in space (which the apple covers) is too short that its path in three-dimensional space is also straight.

This is everything I just described:

Gravity is not a force like others but is a consequence of the '**curvature of space-time**'. The presence of mass and energy curves space-time.

'Curving' of space-time means that near a mass, space contracts and time dilates and this effect goes on decreasing as we move away from the source of mass (or energy).

All the objects follow the shortest path, that is, a straight path in four-dimensional space-time. These paths are known as '**geodesics**' and space pushes the objects to follow these geodesics (straight paths). However, the objects may seem to move in curved paths in three-dimensional space.

6.3 Gravitational Waves

We often represent space-time two-dimensionally as a 'fabric' (Fig 6.3).

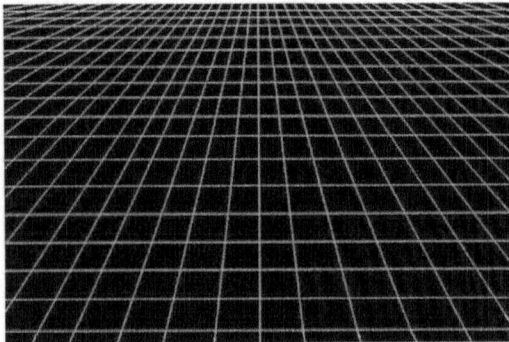

Fig 6.3 (above) - Two-dimensional representation of four-dimensional spacetime

The distortion in space-time 'fabric' is represented in Fig 6.4.

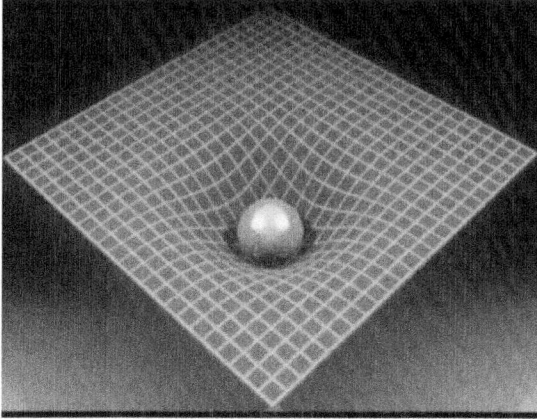

Fig 6.4 (above) - Distortion in four-dimensional spacetime

Space-time can also be represented three-dimensionally and distortion in it is represented in Fig 6.5.

Fig 6.5 (above) - Distortion in four-dimensional spacetime (3-dimensional representation)

Space-time can be 'disturbed' and these disturbances in it are called '**Gravitational Waves**'. A wave can be defined as a disturbance in anything (like a medium or field).

The waves which disturb a medium [like water waves are disturbances in the water, sound waves disturb all solid, liquid and gas] are called '**Mechanical Waves**'.

The waves which disturb the electromagnetic field, like light waves, are called '**Electromagnetic Waves**'.

Similarly, gravitational waves disturb space-time. Gravitational waves are represented in Fig 6.6.

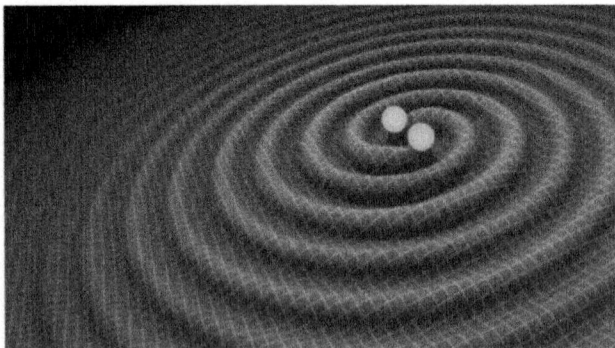

Fig 6.6 (above) - Gravitational waves caused by collision of two 'Neutron Stars'

It can be observed in Fig 6.6 that how 'ripples' are created in the 'fabric of space-time' as are created in the water when a pebble is dropped in it. A thing to be clarified is that this is JUST a representation.

In chapter 4, it was described that if the Sun disappears, it will take eight minutes to know about the darkness because light takes eight minutes to travel from the Sun to Earth.

Gravitational waves also travel at the speed of light. This means that Earth will take eight minutes to leave its orbit (as gravity travels at light's speed). Newton's theory predicts that if the Sun disappears, all the planets will instantly leave their orbit, but this would mean gravity travels faster than light, which violates the special theory of relativity. But, as gravitational waves travel at a finite speed, it will take some time for the planets to leave their orbit (Fig 6.7).

Another thing I want to clarify is that the speed of gravitational waves is also the same for every observer because they travel at the speed of light. Anything which travels at the speed of light will have the same speed for every observer.

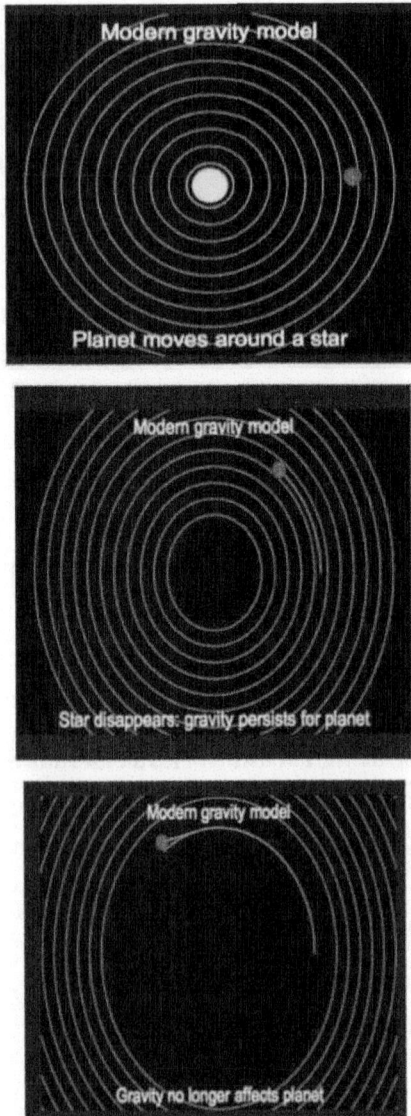

Fig 6.7 (above) -

7

THE UNCERTAINTY PRINCIPLE AND QUANTUM PHYSICS

7.1 Introduction

Einstein's special theory of relativity is not only applicable to the objects moving with very high velocities, but we can also apply special relativity's equations in daily life situations. But, in this case, the results are (almost) similar to that of classical Physics. In other words, the equations of relativity, when applied to daily life situations, reduce to the equations of classical Physics. But, we still use classical Physics due to its simplicity.

Classical Physics isn't applicable to very high-speed situations. But, there is another problem with classical Physics, it is not applicable to the '**small** scale **of the universe**'. It isn't applicable to the atomic scale. This problem is not only with classical Physics but also with the general theory of relativity. But, the special theory of relativity works fine on the small-scale.

The 'Physics of the smallest' is '**Quantum Physics**' or '**Quantum Mechanics**', or simply **QM**. Actually, quantum Physics was first introduced in 1900. The idea of QM started five years before the idea of relativity. Einstein himself had a very important role in QM. He also won the Nobel prize in Physics in 1921 for his work on QM in 1905 but he never won a Nobel prize for his greatest theories, **GR** (general relativity) and **SR** (special relativity).

The foundation of quantum Physics was laid by mainly two (German) people, Max Planck and Albert Einstein.

7.2 Need for Quantum Physics

To understand why QM was introduced, we should discuss a little bit about '**Waves**'.

A wave is a disturbance in a '**Medium**' (like air or water) or some '**Field**' (like a gravitational field, electromagnetic field etc.), we've already discussed that many times.

But, how can a medium or field be disturbed? The answer is simple — with continuous transfer of energy.

So, in other words, a wave is a transfer of energy from one place to another in a medium or field. A familiar example of a wave is a water wave which is produced when you drop anything in the water.

A wave has '**Crests**' (peaks) and '**Troughs**' (bottom point); it moves in 'cycles', that is, the medium or field (which is 'disturbed' by the wave) repeatedly moves up and down (for example water waves or light waves) or left and right (for example, sound waves). This means that the oscillating quantity in water waves is water; in light waves it is the electromagnetic field and in sound waves it is the medium in which the sound wave is travelling. Now there are some terms associated with waves, we are going to discuss them now.

The distance covered by a wave in one single cycle is called its '**Wavelength**'. Wavelength is also equal to the distance between two successive crests or troughs.

Another quantity associated with a wave is its '**Frequency**'. Frequency is the number of cycles or '**Oscillations**' completed by a wave in a single second.

One '**Cycle**' or '**Oscillation**' means moving to the extreme top (or right) position from the original position and then to the extreme bottom (or left) position and again back to the original position.

'**Time Period**' is the time taken by the wave to complete one cycle. It can be concluded now that frequency is just the inverse of the time period (as frequency is cycles completed in one second and time period is the time taken to complete one cycle):

Frequency = 1 / Time Period

And:

Time period = 1 / Frequency

A wave is illustrated below in Fig 7.1.

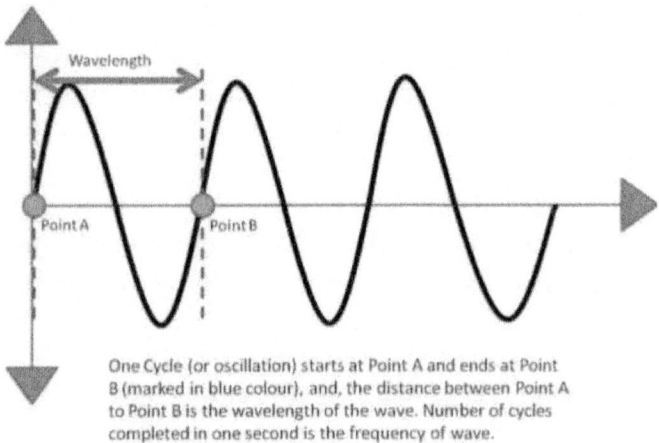

Wavelength

Point A

Point B

One Cycle (or oscillation) starts at Point A and ends at Point B (marked in blue colour), and, the distance between Point A to Point B is the wavelength of the wave. Number of cycles completed in one second is the frequency of wave.

Fig 7.1 (above) - A Wave

According to the '(classical) Electromagnetic Wave Theory' (developed by James Clerk Maxwell), light is also a wave. It disturbs the electromagnetic field. So, light is an electromagnetic wave or electromagnetic radiation. According to this theory, light is emitted and absorbed 'continuously' (in the form of waves). It also states that the 'Intensity' (or brightness) of light is directly proportional to the energy carried by the light waves, and, frequency and wavelength of light are independent of the energy. Actually, light's frequency and wavelength decide its colour. Every different colour has a different frequency (and wavelength).

The word '**Electromagnetic Radiation or Electromagnetic Wave**' is more correct than light because '**Light**' generally refers to the electromagnetic radiation of visible range, that is, the electromagnetic radiation that our "eyes can see". This frequency ranges from the colour red (**Frequency = 43 × 10^{13} Hertz**) to violet colour (**Frequency = 75 × 10^{13} Hertz**). The visible range of light contains seven colours — Red (Lowest Frequency), Yellow, Orange, Green, Blue, Indigo and Violet (Highest Frequency). The above list is in the order of increasing frequency (and decreasing wavelength).

Table 7.1 Shows the frequencies and wavelengths of all the electromagnetic radiation.

Type of Radiation	Wavelength (in nanometres) 1 nanometer = 10^{-9} metres	Frequency (in Hertz) 1 Hertz = 1 cycle or oscillation per second
Gamma Rays	Smaller than **one nanometre**	Greater than **3 × 10^{17}**
X Rays	**1 to 10**	**3 × 10^{16} to 3 × 10^{17}**

Continued…

Ultraviolet Radiation (UV Rays)	10 to 400	75×10^{13} to 3×10^{16}
Visible Light (or simply 'Light')	400 to 700	43×10^{13} to 75×10^{13}
Infrared Radiation	700 to 100000 (or 10^5)	3×10^{12} to 43×10^{13}
Microwave Radiation	10^5 to 10^8	3×10^9 to 3×10^{12}
Radio Waves	Greater than 10^8	Smaller than 3×10^9
TABLE 7.1 (Above)		

It can be easily observed from Table 7.1 that the wavelength of light varies inversely to its frequency.

The speed or velocity of a wave is equal to its frequency multiplied by its wavelength. We can easily derive this relation. We know that velocity is distance travelled by time taken. So, velocity of a wave is distance travelled by the wave divided by the time to cover that distance. And we know that in a time equals the time period of the wave, the wave travels a distance which is equal to its wavelength (in one cycle it covers a distance equal to its wavelength and takes time equal to its time period). So:

Velocity of a Wave = Wavelength/Time Period

And we know that time period is just the inverse of frequency of the wave. So:

Velocity of a wave = Frequency of the wave ×
Wavelength of the wave

Obviously, frequency of a wave is then the velocity of the wave divided by its wavelength and the wavelength of a wave is the velocity of the wave divided by its frequency.

But, the electromagnetic wave theory fails to explain some natural phenomena of light, namely, the **'Black Body Radiation'** and the **'Photoelectric Effect'**.

7.2.1 Black Body Radiation

A **'Black Body'** is an object, which can *perfectly* absorb and radiate (emit) electromagnetic radiation (light) of any frequency. The electromagnetic radiation it emits is known as the **'Black Body Radiation'**.

When such an object, for example, an iron rod is heated, it firstly becomes red, on increasing the temperature becomes yellow and, on constantly increasing the temperature, the frequency of radiation increases and the rod finally starts to glow with blue light.

It should be noted that an increase in temperature means an increase in the (heat) energy.

According to the electromagnetic wave theory, the frequency of radiation is independent of the energy. But, the frequency of radiation increases (in the case of black body radiation) with the increase in energy. So, electromagnetic wave theory is unable to explain the black body radiation.

7.2.2 Photoelectric Effect

'**Photoelectric Effect**' is the emission of 'electrons' (constituent particles of atoms) when light rays strike a metal surface. The emitted electrons are known as '**Photoelectrons**'. This effect was first discovered by Heinrich Hertz, a German Physicist.

In metals, a number of 'free electrons' are there, that is, the electrons which 'feel' very less attraction to the atom's nucleus. As the free electrons are also (very weakly) bound to the nucleus, they require some minimum energy to leave the surface of metals because the 'positively charged' nucleus is continuously attracting the 'negatively charged' electrons.

But, it was observed that electrons are only ejected if the frequency of light (striking the metal surface) is greater than a particular 'minimum frequency', known as the '**Threshold Frequency**' of a metal, that is, the ejection of electrons depends upon the frequency of light only. This means the emission of electrons has nothing to do with the intensity of light.

But, according to the electromagnetic wave theory, not frequency but intensity "decides" the energy of the light rays. So, electromagnetic wave theory is unable to explain this observation because according to this theory, only a minimum intensity should decide whether the electrons will be ejected or not.

The photoelectric effect is illustrated in Fig 7.2.

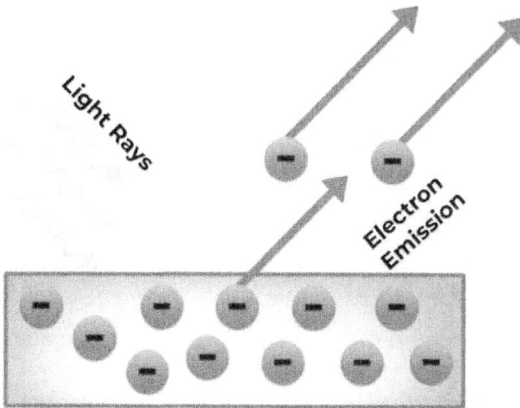

Fig 7.2 (above) - The Photoelectric Effect

7.2.3 Planck's Quantum Theory and the Dual Nature of Light

To explain the black body radiation, a German Physicist, Max Planck suggested that light is not emitted continuously in the form of waves, but, 'discontinuously' in some small 'packets' of energy known as '**Quanta**' (singular of quanta is 'quantum'). A '**Quantum**' can be defined as the most fundamental unit of anything. One quantum of light is known as a '**Photon**'. This means that light is made of particles known as photons.

Does this mean that light is not a wave? Actually, both things are correct. Light shows a 'dual' nature of both waves and particles. Each photon is associated with a wave. There are many phenomena of light which can only be explained by wave theory (like Interference) and not by particle nature, and some can only be explained by particle nature.

Planck also suggested that the energy of a photon is directly proportional to the frequency of the corresponding light wave. That is, the greater the energy of the photon, the greater the frequency of the corresponding light wave, and vice versa. And, the relation between energy and frequency is given by:

Energy of a Photon = Planck's Constant ×
Frequency of corresponding Light Wave

The above relation is also known as the '**Einstein-Planck Relation**'.

The numerical value (value without unit) of Planck's constant is **6.6 × 10^{-34}**.

If we put the value of frequency (in terms of velocity and wavelength) in the above relation, we get:

Energy of a Photon = Planck's constant × (Velocity of the Light / Wavelength of the corresponding light wave)

Now, as the energy is independent of intensity, the black body radiation can be explained as follows:

When an object, like an iron rod, is heated, it starts to glow with red light first. As we increase the temperature, the energy increases, and hence, the frequency increases. So, the iron rod starts to glow with a light of higher frequency. The frequency of the radiation emitted will increase with temperature.

Using Planck's Quantum Theory, the photoelectric effect can be explained as follows:

We know that free electrons require a minimum energy to leave the surface of the metal. And, as frequency and energy are directly proportional, light of a minimum frequency is required to eject electrons from the metal.

As we strike light on the metal, the photons will strike the electrons and they will give their energy to the electrons and electrons will be ejected as they got the minimum energy.

7.3 Wave Particle Duality

Max Planck and Albert Einstein laid the foundation for the development of Modern Physics.

But, the dual nature described in Planck's Quantum theory was only about photons and electromagnetic waves.

In 1924, a French physicist, Louis de Broglie suggested that not only photons but the particles of matter (electrons, quarks and neutrinos, mainly) also exhibit dual nature of particles and waves. This hypothesis of de Broglie was a 'generalization' of Planck's theory. So, the Einstein-Planck relation can be applied to any particle. The wave associated with a matter particle is known as a 'matter wave'.

So, Einstein-Planck relation can be written as:

Energy of a matter particle = Planck's constant × Frequency of the corresponding matter wave

When we will do some math on the Einstein-Planck relation and the Einstein Mass-Energy relation, we will get the following relation:

Wavelength of a matter wave = Planck's Constant / Momentum of the corresponding matter particle

(The momentum of a particle is the mass of the particle multiplied by its velocity.)

The above relation is also called '**de Broglie's Wavelength Relation**'. The relation described is applicable to any type of object.

So, one can now ask what the 'oscillating stuff' in matter waves is; what is the field or medium which is being disturbed by matter waves? It is actually the '**Wave Function**'. So, what is the wave function now? Wave function is a (mathematical) quantity which describes the quantum state of any system. We'll discuss some more things about the wave function later in this chapter.

But, if matter shows dual nature, why can't we observe the wave nature of matter in our daily life.

So, take an example of any bullet of four grams moving with a velocity of **400 metres per second**. Consider its diameter is **five centimetres**. The mass of the bullet in kilograms will be **0.004 kg**.

So, its momentum will be **0.004** multiplied by **400**, that is, **1.6 kg m/s** (mass × velocity). Now, we apply the de Broglie relation, and obtain that the wavelength of the corresponding wave will be only about **4 × 10^{-34} meters** and the diameter of the bullet is five centimetres or **0.05 meters**, which is much larger than its wavelength. So, here, absolutely, particle nature dominates by a large factor.

So, for objects far larger than an atom, particle nature dominates due to having large size and very short wavelength.

This dual nature of particles and waves is called the '**Wave-Particle Duality**'. We know that Einstein proved the dual nature of light (by explaining the photoelectric effect), but what about other particles, how can the de Broglie hypothesis be proved? It can be proved with the '**Two-Slit Experimen**t' or the '**Double Slit Experiment**'.

7.3.1 Double Slit Experiment

To understand what the double slit experiment is, we should first discuss something about 'Interference' of waves. Interference can be defined as 'interaction of waves'. This interaction can be 'constructive' or 'destructive'.

When the waves are in the '**same phase**', that is, when peaks (or crests) of two waves 'meet', the two waves combine to become a single, but stronger wave, and, when the waves are '**out of phase**', that is, when the peak of one wave meets with the bottom (or trough) of the second wave, the two waves cancel out each other.

Interference is illustrated in Fig 7.3.

Constructive Interaction - Same Phase

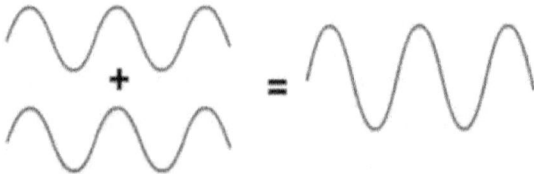

Destructive Interaction - Out of Phase

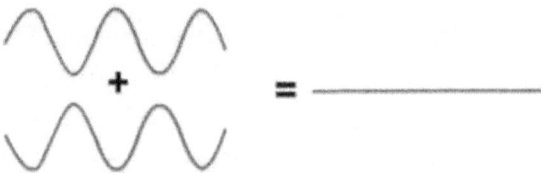

<u>Fig 7.3 (above) -</u> Interference of Waves.

So, now we can finally come to the two-slit or double slit experiment.

Double slit experiment was initially performed by a British polymath and physician (not Physicist), to prove that light is a wave and he did it successfully. He was the first to prove that light is a wave (even earlier than Maxwell).

Setup of the double-slit experiment: A light source, single slit, double slit and a screen were taken.

The single slit was taken to make the light pass through the double slit (actually when light passes through a small opening, like this slit, it bends. This bending of light when it passes through such a slit is called '**Diffraction**'). The light wave splits into two different waves when it passes through the double slit. The setup is illustrated in Fig 7.4.

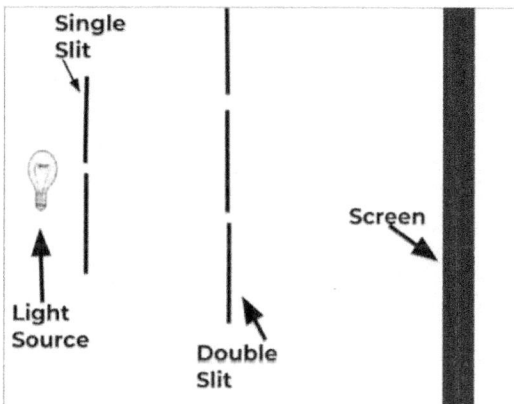

Fig 7.4 (above) - Setup of Double Slit Experiment.

Observations: When a single light wave will split into two different waves, the two waves will interfere with each other and will create an 'Interference Pattern' illustrated in Fig 7.5.

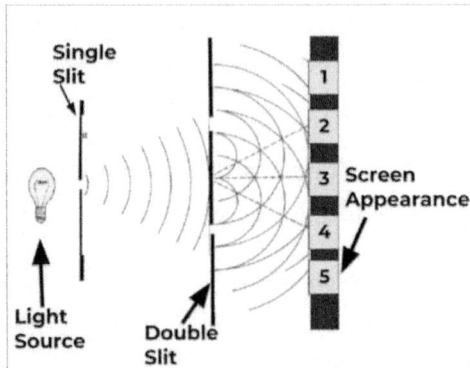

Fig 7.5 (above) - Interference Pattern obtained in Double Slit Experiment.

In Fig 7.5, light appears on the light-coloured 'spots', that is, at the positions **1**, **2**, **3**, **4**, and **5** due to constructive interference and dark spots are formed due to destructive interference where waves cancel each other and light doesn't appear on these spots.

Conclusion of Two-slit Experiment: Young's two-slit experiment concludes that light is a wave because it interferes (and diffracts). If light showed only particle nature, then the light would appear only at the positions **2** and **4** in Fig 7.5.

This surely doesn't prove the de Broglie hypothesis, but it can! According to de Broglie hypothesis, matter particles also behave like waves, so let's take an example of an electron (a matter particle).

To prove the de Broglie hypothesis, the setup will remain the same, only the light source will be replaced by an 'Electron Gun', which will fire electrons and the screen will be replaced by a 'Detector' which will detect the electrons. This experiment was suggested by an American physicist, Richard Feynman and this was called 'Feynman's thought experiment'. When this experiment was actually performed, the result was the same as Feynman suggested and this led to the ultimate proof of wave-particle duality and the de Broglie hypothesis.

When this experiment was performed using a single electron, there was still an interference pattern on the detector, so this means that the electron would be passing through both the slits at the same time and interfering with itself! This can actually happen because a single electron also shows wave nature; the 'electron wave' will split into two waves and then interfere with itself, giving an interference pattern.

The double slit experiment using electrons is illustrated in Fig 7.6.

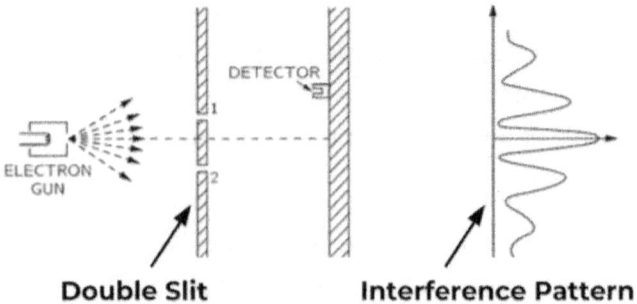

Double Slit   **Interference Pattern**

Fig 7.6 (above) - Interference Pattern obtained in double slit experiment of electrons.

7.4 Heisenberg's Uncertainty Principle

In 1814, a French mathematician Marquis de Laplace introduced the 'Principle of Determinism' in which he suggested that if we know the present state of the universe, then, using the correct laws of Physics, we can find the future or past state of the universe with 100 percent certainty.

It seems to be true because if we know the present position and momentum of Earth, Sun and our Moon, we can find out when the next solar or lunar eclipse will take place by using Newton's laws. It seems that we can predict any motion etc. like this. Everyone was satisfied with this principle of determinism.

After developing his general theory of relativity and being awarded by the Nobel prize for the photoelectric effect in 1921, Einstein was doing research in quantum theory, but his research was interrupted by (another) German physicist, Werner Heisenberg in 1928 when Heisenberg presented his 'Uncertainty Principle'.

It was the most important result of Planck's theory that was realized only in 1928, by Heisenberg.

This principle states that we can't measure the position and momentum of a particle with 100 percent certainty. If we increase the certainty in one of these two, the certainty in the other quantity decreases.

There is a mathematical relation between uncertainty in position, momentum and Planck's constant, which is:

Uncertainty in Position × Uncertainty in Momentum = Always greater than Reduced Planck's Constant / 2

(Numerical value of the reduced Planck's Constant is **1.0545718 × 10^{-34}**).

But, then why can we find out (with certainty) when the next solar or lunar eclipse will take place using Newton's laws?

Actually, there is uncertainty in the position and momentum of a particle because it also shows wave nature. If this is true, then obviously we can't measure both the position and momentum of a particle precisely, because a wave is a transfer of energy in a medium or a field. As a wave is nothing but a transfer of energy, so we can't give a definite position to a wave. A wave is 'delocalized'. And because all particles also behave as waves, so, now particles don't have any absolute location or position or momentum, they are delocalized as well.

But, then why doesn't this apply to objects far larger than atoms. This is because (as we discussed above) the size of large objects is much larger than the wavelength of their corresponding matter wave, so the uncertainty in their position and momentum is almost equal to zero.

So, now one thing we know is that in quantum physics, we mainly have to work with probabilities, because nothing is certain now. These probabilities are described by the wave function. The (square of) wave function describes all the possible positions (and momenta) at which a particle can be found at some time and also gives the different probabilities for these positions (and momenta).

7.5 Elementary Particles and Fundamental Forces of Nature

Suppose you have an object, like a wooden box. You cut the wooden box into two pieces. Then you cut one of the pieces, and you do it continuously, cutting it into smaller and smaller parts. A stage will ultimately come when you will not be able to cut it anymore.

This indivisible 'object' or particle is called an 'Elementary Particle'. Elementary particles are indivisible, in other words, elementary particles are quanta like a photon (and photon is also an elementary particle).

This example we took above was a 'thought experiment' of many Greeks thinkers in around the fifth century BC. Initially, by the Greeks, the indivisible particle was called 'Atom'.

But it was only in 1805 that an English Physicist and chemist, John Dalton put forward a complete description of the atom in his atomic theory. The following were the main postulates of 'Dalton's atomic theory':

1. Matter (anything which occupies space and has mass) is made up of very tiny particles called 'Atoms'.

2. Atoms are indivisible and can't be destroyed.

3. All atoms of a particular element have the same mass.

4. Atoms of different elements have different masses.

5. Atoms of different (or even same) elements combine with other atoms to form new molecules of elements as well as molecules of compounds.

But, there are some failures of this theory. Firstly, atoms are divisible. The most fundamental particles of atoms are '**Quarks**' and '**Electrons**'.

Atoms of the same element can have different masses, like, hydrogen. A hydrogen atom is found in three different 'states', in which every state has a different mass.

The first state is '**Protium**' (normal hydrogen), the second is '**Deuterium**' (heavy hydrogen) and the third one is '**Tritium**' (further heavy hydrogen). These different states of the same element whose atoms differ in mass are called '**Isotopes**'.

Atoms of different elements can have the same mass, for example, one atom of a gaseous element '**Argon**' and a metal '**Calcium**' have the same mass.

These types of elements which have the same mass but differ in other aspects are called '**Isobars**'.

7.5.1 Modern Picture of Atom

In 1896, JJ Thomson, a British Physicist, discovered '**Electrons**' in an atom. We now know that electrons are elementary (indivisible) particles. An electron has very small mass and size. It is a negatively charged particle and possesses 'one unit of negative elementary charge'.

In 1886, Eugen Goldstein, a German physicist, discovered a positive charge in an atom. It was in 1911, when a student of JJ Thomson, Ernest Rutherford, carried out an experiment and suggested that this positively charged 'thing' is located in the middle of an atom. He named the positively charged stuff as the 'Nucleus' of the atom. Rutherford also suggested that this nucleus consists of some positively charged particles and some zero-charged (neutral) particles. Rutherford named the positively charged particles as 'Protons' in 1919. A proton is not an elementary particle. It is made up of 'Quarks'. Also, protons have 'one unit of positive elementary charge'. The mass of a proton is almost two thousand times the mass of an electron.

Rutherford in his 'Alpha Particle Scattering' experiment in 1911, took a gold sheet, and fired a beam of positively charged particles called 'Alpha Particles'. We know that objects with similar charges (like positive and positive, and negative and negative) repel each other (and with different charges attract each other). When the experiment was carried out, many alpha particles went straight through the atom, some were deflected by small angles and some were turned back completely.

This 'deflection' in the beam of alpha particles was due to some positive charged 'stuff'. On the basis of these observations, Rutherford put forward his model of the atom which is as follows:

1. Most of the atom is an empty space (because most of the alpha particles went straight).

2. At the centre of the atom, there are positively charged protons and some other particles with zero charge (neutral). The centre of the atom [containing proton(s) and the neutral particle] is known as the '**Nucleus**' of the atom. So, the nucleus as a whole is positively charged.

3. Almost all the mass of an atom is concentrated in the nucleus.

4. Electrons orbit the nucleus due to opposite charges (negative and positive).

5. Atom has a total of zero charge because the number of protons is equal to the number of electrons in a particular atom.

The zero charged (neutral) particles in an atom are called '**Neutrons**' and were discovered by a student of Rutherford, James Chadwick in 1932 (JJ Thomson discovered electrons, his student, Rutherford named protons, and Rutherford's student James Chadwick discovered neutrons). The mass of a neutron is slightly higher than a proton. Neutrons (like protons) are also made up of quarks.

When an electron is forced to 'penetrate' into a proton, a neutron and a '**Neutrino**' is formed. Neutrinos, unlike protons and neutrons, are elementary (indivisible) particles like electrons and quarks (not to be confused with 'Neutrino' and 'Neutron').

The constituents of an atomic nucleus are collectively known as '**Nucleons**', that is, protons and neutrons are also called nucleons (collectively).

Also, any particle made up of quarks is known as a 'Hadron', like protons and neutrons.

Like Dalton's theory, Rutherford's model of the atom also has some limitations. Firstly, it fails to explain in which type of orbits electrons revolve around the nucleus and, secondly, it fails to explain why electrons don't fall into the nucleus if the electron is orbiting the nucleus. Because, when anything moves in circular or elliptical orbit or changes its direction with time, it is considered as an accelerated motion, because acceleration not only depends on the change in speed but also on the direction of motion (as discussed in chapter 1).

As direction changes, acceleration takes place. Now as to accelerate something, some force is needed and energy is used up in applying a force, so electrons should fall into the nucleus after some time. This reason is actually not very precise. This is the more correct reason: an accelerated charged particle radiates (loses) energy in the form of electromagnetic waves. This is because of the fact that a charged particle creates an electromagnetic field around it. As it accelerates, it creates disturbances in the electromagnetic field, and, these disturbances are, of course, electromagnetic waves.

The same situation is with the electrons accelerating around the nucleus in circular orbits. So, now, electrons should also lose energy in the form of photons or electromagnetic waves/radiation and then finally fall into the nucleus (Fig 7.7).

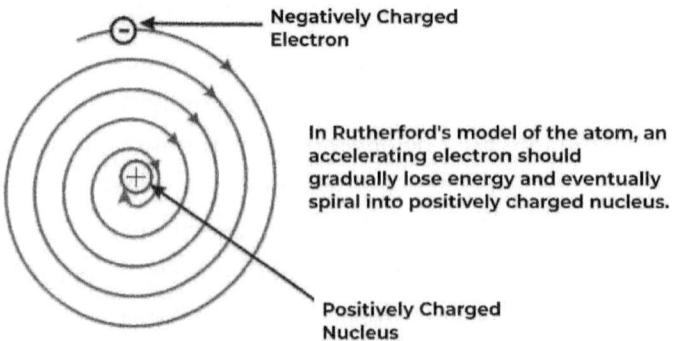

Negatively Charged Electron

In Rutherford's model of the atom, an accelerating electron should gradually lose energy and eventually spiral into positively charged nucleus.

Positively Charged Nucleus

Fig 7.7 (above) - Electron falling into Nucleus

This problem was solved by another Physicist, Niels Bohr, in 1913, when he put forward his model of the atom. Bohr's model of the atom is as follows:

1. Electrons revolve around the nucleus in 'fixed' orbits.

   These orbits are known as '**Energy Levels**' or '**Energy Shells**'. Every energy level (shell) is located at a fixed distance from the nucleus and each shell is represented by **K, L, M, N**… There are integers assigned to these shells starting from **1, 2, 3, 4**… These integers are known as '**Principal Quantum Number**'.

   The electron(s) present in the shell nearest to the nucleus (**K shell**) has the lowest energy and this energy is called '**Ground State Energy**' and the energy (of an electron) increases as the distance from the nucleus increases. The energy of electrons in shells, **L, M, N** or any higher energy level is called '**Excited State**' (relative to '**K**' shell).

2. The electrons are arranged in these shells in a very special way. These electrons don't lose energy when in a particular energy shell.

   When it drops down in a lower energy shell, it radiates energy in the form of electromagnetic radiation (photons). If an electron raises up to a higher energy shell, it absorbs energy in the form of electromagnetic radiation.

We can easily find the amount of energy released or absorbed using the Einstein-Planck relation as follows:

Energy Absorbed or Released = Planck's Constant × Frequency of electromagnetic radiation absorbed or released

Bohr's model of the atom is illustrated in Fig 7.8.

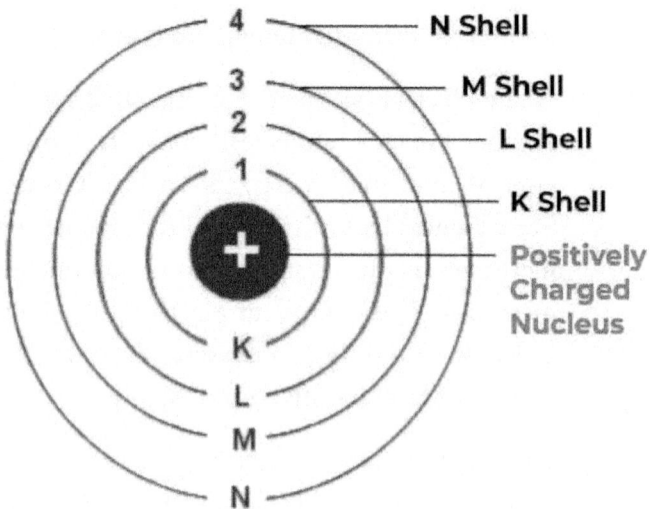

Fig 7.8 (above) - Bohr's Model of Atom

But, the question is — why don't electrons lose energy in a particular shell continuously?

This is because they can only have some fixed values of energies. This means that their energies have discrete values. Electrons can lose energy but not continuously (means discontinuously, in the form of photons not continuously like waves).

This concept is known as '**Quantization**'. When something is discrete, (that is, it has only some fixed values and doesn't change continuously) we say that it is quantized.

To understand quantization more precisely, we can also take an example from daily life. When you travel in a vehicle (like a car) you observe that the speedometer of the car moves continuously as the speed of the car increases and can have any arbitrary value depending upon the speed and acceleration.

On the other hand, you may have observed that the 'fare meter' of the taxi moves discontinuously in the multiples of one dollar etc. This is due to the fact that the fare meter is designed in such a way that it can show only some fixed values in integral multiples of one dollar (one, two, three etc.).

So, we can say that the fare meter is quantized but the speedometer is not.

Now, we should come back to the question, why don't electrons lose energy continuously when in a particular energy level and also why are the orbits/shells located at only some fixed distances from the nucleus?

Actually, according to the wave-particle duality, particles behave like waves. So, we can think of electrons as waves when they orbit the nucleus. Electrons can be found only at those distances (from the nucleus) where constructive interference of various wavelengths of an electron takes place, so that the waves get stronger. This is the reason for the fact that electrons are found only at some particular distances from the nucleus and hence have fixed energies in particular shells. Electrons will not be found at the places where destructive interference of electron waves takes place. The (potential) energy of an electron in an atom depends on its distance from the nucleus. The greater the distance, greater the potential energy of the electron. (Actually, greater the distance, less negative the (potential) energy of the electron, but that's just the same thing.)

This model of the atom is our current picture of atom and is known as the '**Quantum Mechanical model of the atom**', '**Quantum Physical model of the atom**' or the '**Wave Mechanical model of the atom**'.

Another postulate of the quantum physical model of the atom is that the position (and momentum) of the electrons can't be determined with 100% probability, there will be an uncertainty (according to the uncertainty principle).

So, this section 7.5.1 was all about how our understanding of the atom increased with time. The next section, 7.5.2 is about the various elementary particles, that is, the fundamental, indivisible particles and the forces by which they interact.

7.5.2 Standard Model of Particle Physics and the Fundamental Forces

So, now we can finally discuss the various (discovered) elementary particles and the fundamental forces of nature.

We have already discussed the term 'elementary particle'. These are the most fundamental particles currently known. It will not be a surprise if more fundamental particles will be discovered in the future.

In this section, firstly we will explore the '**Standard Model of Particle Physics**' and the various discovered elementary particles, and, the four forces which shape our universe.

The standard model is a theory describing three of the four fundamental forces and it also classifies the elementary particles.

The four fundamental forces which shape our universe are —

1. Strong Nuclear Force.

2. Weak Nuclear Force.

3. Electromagnetic Force.

4. Gravity (Not explained by the Standard Model, explained by general relativity).

The discovered elementary particles are —

1. Quarks (up, down, top, bottom, strange, charm).

2. Leptons [electrons, muons, tau, neutrinos (electron neutrinos, muon neutrinos, tau neutrinos)].

3. Bosons (gluons, photons, Higgs boson, W bosons, Z boson).

All this is in Fig 7.9.

FERMIONS (Matter Particles)				BOSONS (Force Carriers)	
QUARKS	Up (u)	Charm (c)	Top (t)	BOSONS	Gluon (g)
	Down (d)	Strange (s)	Bottom (b)		Photon (γ)
					Z Boson (Z)
LEPTONS	Electron (e)	Muon (μ)	Tau (τ)		W Bosons (W)
	Electron Neutrino (v_e)	Muon Neutrino (v_μ)	Tau Neutrino (v_τ)		Higgs Boson (H)

Fig 7.9 (above) - Standard Model of Particle Physics with names and symbols of elementary particles

Also, for every particle there exists a corresponding 'Anti-particle'. When a particle and its antiparticle come in physical contact with each other, they 'vanish' converting into 'energy'.

So, there are a total of seventeen elementary particles and their corresponding antiparticles which 'make' the universe (there are 'Dark Matter' and 'Dark Energy' also) and four forces which shape the universe.

There are some other particles like gravitons and sterile neutrinos which are still hypothetical. Sterile neutrinos are supposed to be elementary particles of dark matter. Dark Matter and Dark Energy are 'mysterious' types of matter and energy which will be discussed in the next chapter.

The particles in the standard model can be broadly classified into 'Fermions' and 'Bosons'.

Fermions are the matter particles including all types of quarks and leptons. These particles interact with each other through the forces mentioned above.

The second, bosons, are responsible for causing these fundamental forces, that is, they are 'Force-Carrying Particles'.

The exchange of bosons between two or more fermions causes the fermion(s) to attract or repel each other.

To a rough approximation, it can be understood with Newton's third law of motion (conservation of momentum, more precisely). When a fermion 'releases' a boson, it is pushed backwards by the recoil, then, the boson is 'absorbed' by a second particle, and, due to this collision, the particle is pushed forward. This is illustrated in Fig 7.10.

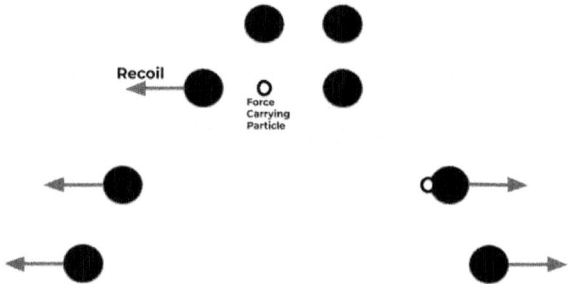

Fig 7.10 (above) - Mechanism of Force Carrying Particles (Bosons)

7.5.2.1 The Strong Nuclear Force

So, now we are going to talk about '**Quarks**', '**Gluons**' and '**Strong Nuclear Force**' or simply '**Strong Force**'.

There are six types (flavours) of quarks — **up**, **down, charm, strange, top** and **bottom**. Every type of quark comes in three '**Colours**' — **Red, Blue** and **Green** (Actually they don't possess any 'real' colour because they can't be seen as their size is much smaller than the wavelength of visible light).

The nucleons of an atom, that is, protons and neutrons are made up of three quarks each. Protons are made up of two 'up quarks' and one 'down quark'. As the charge on an up quark is positive (two-thirds unit of elementary positive charge) and that of a down quark is negative (one-third unit of elementary negative charge), so the proton, as a whole, is one unit of elementary positive charge.

Similarly, a neutron is neutral, that is, it doesn't possess any charge This is because of the fact that neutrons are made up of two down quarks and one up quark. The total charge of two down and one up quark becomes zero.

Moreover, quarks possess a property called **'Colour Charge'** which makes them 'feel' the strong force. The strong force is 'carried' by 'gluons'.

Quarks and Antiquarks are the particles which interact through the strong force. Gluons themselves also interact with the strong force. Another thing about strong force is that it only acts in a very short range which is about 10^{-15} **meters**. It is a very short ranged force because gluons are comparatively more massive than other bosons. Within its range, strong nuclear force is the strongest of all the fundamental forces of nature.

Quarks also make up a particle called **'Meson'**. A meson is composed of one quark and an antiquark bound by the strong nuclear force. So, now the question arises, why don't mesons annihilate if they are composed of one particle (quark) and its anti-particle (antiquark).

The reason is that the quark and antiquark aren't of the same 'flavour'. For example, if the quark in the meson is an 'up quark' then the antiquark is 'down antiquark', so they won't annihilate. This is the simple reason for the existence of mesons. In this way, there are a total of ten types of mesons and their ten 'anti-mesons'.

Another thing (also described previously) is that the type of particles made up of quarks and antiquarks are known as 'hadrons', so mesons are also hadrons.

7.5.2.2 The Weak Nuclear Force

Now we are going to talk about 'W and Z Bosons' and the 'Weak Nuclear Force' or simply 'Weak Force'.

The weak force is caused by W and Z bosons. It is responsible for the 'radioactive decay' of particles. For example, in a nuclear chain reaction, the weak force is responsible for the decaying of a neutron into a proton, electron and an anti-neutrino.

The range of this force is very short, shorter than the strong force. It is only about 10^{-18} meters which is even smaller than the 'thickness' of a proton. It is the third strongest fundamental force.

7.5.2.3 The Electromagnetic Force

In this section, we will explore '**Photons**' and the '**Electromagnetic Force**' or simply 'Electromagnetism'.

Whether it is the attraction or repulsion of magnets or flowing of current or any other electrical (or magnetic) phenomenon, it is electromagnetism which is responsible for this. And, electromagnetism is caused by 'Photons' which are the quanta of light. So, in other words, light causes electromagnetism.

We know that light is an electromagnetic wave. It means that light creates 'ripples' in electric and magnetic fields.

A photon is a very different kind of particle. Unlike others, it doesn't decay because of its speed (as described in chapter 3). According to special relativity, time doesn't pass for a photon, and that is why it doesn't decay.

The range of electromagnetic force is infinite because photons are almost massless particles, but, as distance increases, its strength decreases. It is the second strongest fundamental force.

A thing which is important to discuss here is that electromagnetism is the most common force we

EXPERIENCE and OBSERVE in our daily lives, even more common than gravity. Whether it is the normal force applied on an object by a surface, the frictional force, the tensional force produced in a string or the restoring force in a spring, it is just electromagnetism working here. All these forces are electromagnetic. Electromagnetism is working inside our bodies. You reading this book—turning pages, seeing the text and interpreting it—all this is nothing but electromagnetism.

7.5.2.4 'The Gravity'

We have discussed gravity almost a thousand times in this book. But, in this chapter, we shall discuss it in a little bit of 'quantum mechanical way'.

We know that there are four fundamental forces of nature, three of which are described by quantum mechanics. The last one, which shouldn't be referred to as a 'real force', is described by Einstein's general theory of relativity (or GR).

Gravity, like electromagnetism, has infinite range, and, as the distance increases, its strength decreases. It is the weakest of all fundamental forces, but, at large masses, it can be even stronger than any other force.

We also know that we apply quantum mechanics at very small scales and general relativity at extremely large scales, comparatively (actually, not at large scale, but at the places of strong gravitational fields).

Now, the problem is that when we try to apply general relativity at quantum scale, it gives us very 'scary' results (which will be described soon), and, when quantum mechanics is applied to strong gravitational fields it gives the same type of scary results.

So, why do we need to apply general relativity to small scales and QM to strong gravitational fields? It is a good question, but there is some stuff which is extremely small in size (almost the smallest) and also possesses a very strong (actually, strongest) gravitational field.

This 'stuff' is known as a '**Singularity**' which is a point in space-time which has infinite density and is (almost) zero in size. Due to this infinite density, the curvature of space-time becomes infinite around the singularity. Such a singularity is also 'found' inside a '**Black Hole**'.

Now, general relativity and quantum mechanics both can't be applied simultaneously, because, as described, they start to give some very 'ridiculous' results when applied with each other. It means both these theories are incompatible with each other.

But, why? There are many reasons for this, for example, general relativity predicts that space-time is continuous (like, a point can have any random position in space-time and it can change in any manner) and not 'discrete' or quantized, so, gravity also travels continuously (in the form of continuous disturbances or curves in space-time). It means space-time and gravity aren't made up of fundamental particles etc. But, loosely speaking, quantum mechanics itself is designed only for 'discrete stuff'.

As we discussed before, general relativity predicts that gravity travels in waves (disturbances or distortions in space-time) but it can't have any particle nature. The other three forces, with no doubt, are quantized. The Strong force is quantized as gluons, weak force as the W and Z bosons, and electromagnetism as photons.

So, this is a major reason that GR and QM are inconsistent with each other.

Another reason is that GR, sometimes, disproves the uncertainty principle which is a very fundamental principle of QM (and the whole Physics in general). According to GR, any point in space-time can be assigned to a certain location. But, according to the uncertainty principle, there should be some uncertainty in this. But, now, this argument is not completely correct. This is because space-time doesn't have mass, and thus, a point (in a geometrical sense) in space-time can have a certain location because the uncertainty principle states that something having mass would 'show' an uncertainty in its position and momentum (anything can only have momentum when it has mass. A point in space-time is not even a 'physical' thing).

Now, there is another question: why don't we just 'throw out' GR (and QM) and quantize gravity (or 'de-quantize' other forces), that is, why don't we simply accept that gravity is carried by 'gravitons' (or try to find different models, similar to general relativity, for other forces as well)?

This is because of the fact that gravitons are not detected yet. Also, GR has been always proved 'correct' in all the experiments and observations.

Actually, neither GR nor QM is wrong, but both theories are 'incomplete' and both are in development yet.

There are many theories about the 'unification' of QM and GR, the most popular of them is '**String Theory**'. But none of these theories is proven yet. They are more like hypotheses only.

The next chapter is all about the creation and evolution of the universe (as described by General Relativity), the life of stars and the possible 'death' of the universe.

8

THE CREATION AND EVOLUTION OF OUR UNIVERSE

8.1 Introduction

Until 1927, almost everyone believed that the universe is static and has always existed, that is, it neither expands nor contracts. It was also believed that the universe didn't have any origin and will never end.

It was only in 1925 when an American astronomer, Edwin Hubble, observed that all the galaxies are moving away from us. How he did it and the significance of it, the creation and evolution of this universe (as predicted by general relativity) will be discussed in this chapter.

This is probably the most important chapter of this book.

8.2 The Expanding Universe

When you look up at the night sky, you see the stars, planets and moon. (Fig 8.1).

Fig 8.1 (above) - The night sky seen by naked eyes.

Almost everyone thought that our universe was static and it has existed forever and would exist forever (including Einstein, and he also published a paper regarding a static universe). But one can argue that if the universe were static, gravity would make the universe collapse into a single point, so, if the universe were static, there must be an '**Anti-gravity**' force which would prevent the universe from collapsing, so, in the equations of general relativity (called '**Einstein's Field Equations**'), Einstein introduced a '**Cosmological Constant**' which represents this anti-gravity force, (the equations of general relativity also include Newton's gravitational constant). The numerical value of the cosmological constant is 11×10^{-53}.

Einstein used it to balance the gravity in such a way that the universe would be static. It was in 1922 that a Russian Physicist, Alexander Friedmann derived some equations (called the '**Friedmann**

Equations') using the equations of general relativity which predicted that the universe must be expanding and it must be 'homogeneous', that is, the universe is uniform, it would look "the same from everywhere".

The first prediction of Friedmann was confirmed in 1927. In 1925, when an American astronomer, Edwin Hubble was observing the night sky with his telescope, which was the best telescope at that time, he observed that there were other galaxies also (Everyone believed until 1925 that only a single galaxy, the Milky Way existed).

Hubble also observed that the light coming from these distant galaxies was shifting slightly towards the red colour. To understand what it means, we should discuss the '**Doppler Effect**'.

The Doppler effect is based on the principle that when a source of sound waves goes away from an observer, the volume of the sound will decrease and when it will come near to the observer, the volume will increase. Actually this effect will take place when there is a relative motion between source and observer.

This is because, when the source will go away from the observer, the wavelength of sound will increase, and thus, the frequency and energy of sound will decrease.

A wave is illustrated in Fig 8.2 (also illustrated in the previous chapter), and the Doppler effect is illustrated in Fig 8.3.

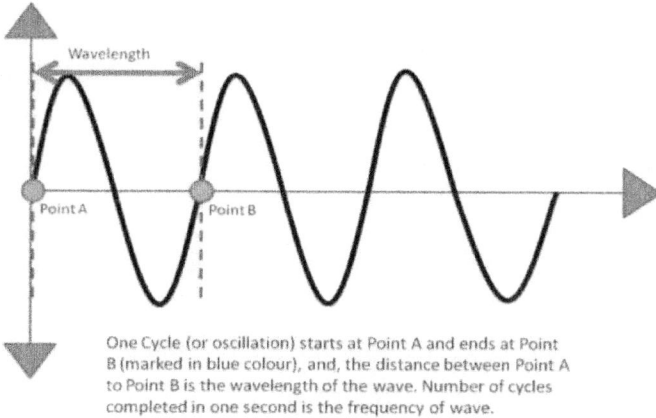

One Cycle (or oscillation) starts at Point A and ends at Point B (marked in blue colour), and, the distance between Point A to Point B is the wavelength of the wave. Number of cycles completed in one second is the frequency of wave.

Fig 8.2 (above) - A wave.

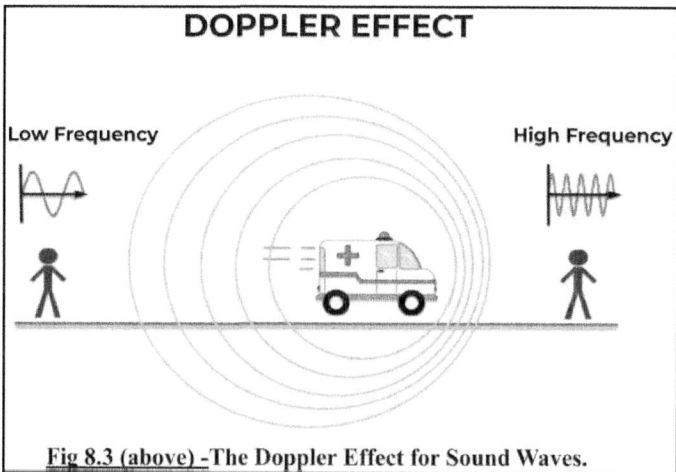

Fig 8.3 (above) - The Doppler Effect for Sound Waves.

This Doppler effect is the same for light waves. When the source of light waves goes away from the observer, frequency of light decreases for the observer and this is called '**Red-shifting of Light**' (because red colour has the least frequency in visible range).

Similarly, when the source comes near the observer, the frequency of light increases for the observer and this is called '**Blue-shifting of Light**'.

Hubble observed that the light coming from distant galaxies is red-shifted. This means that the galaxies are going away from us, and the farther the galaxies are from us, higher their velocities.

Then, in 1927, Georges Lemaitre, a Physicist, suggested the revolutionary idea of an expanding universe. He gave an estimated value of expansion of the universe by dividing the velocity of a galaxy by its distance from earth which was later corrected by Hubble himself and this value is known as the '**Hubble's Constant**' which gives us the value of the rate of expansion of the universe. The value of Hubble's constant is:

20 (km/s)/Mly

'**Mly**' stands for a million light years. This means that if a galaxy is one million light years away from the Earth, it is going away from us with a velocity of

20 km/s. If it is two million light years away (from the Earth), then its velocity is **40 km/s** and so on.

Actually, the red-shifting of light here is not doppler red-shift but '**Cosmological Red-shift**'. The basic thing is the same, the source is going away from the observer, but, in doppler red-shift the source (or observer, or both) moves and in cosmological red-shift, the space (space here refers to the three-dimensional structure) between the source and the observer expands.

This was the confirmation of Friedmann's first assumption of a non-static universe.

His second assumption was that the universe is homogeneous, that is uniform. This means that when you look at a particular volume of the universe at a particular time and again look at the same volume after some billion years, you will find that the pattern is almost the same. It was in 1964 when two American astronomers, Arno Penzias and Robert Wilson, accidentally confirmed the second assumption of Friedmann. Penzias and Wilson were struggling with a signal (noise) received by their radio telescope. They observed that the signal was the same in all directions. This radiation was actually coming from the initial universe.

Actually, light travels at a finite speed, so the stars we see in the night sky are "from the past".

For example, if you see a star five light years away from you, you see it as it was five years earlier. Even when we see the Sun, we see it as it was eight minutes earlier because light takes eight minutes to reach us from the Sun's surface. The same case is with this mysterious radiation. The radiation that Penzias and Wilson obtained was from the initial universe.

The radiation Penzias and Wilson observed was travelling since the universe was 'young' (380,000 years old). When they converted this radiation into visible light, they obtained the 'DNA' of the universe. This 'image' is called the '**Cosmic Microwave Background**' (Fig 8.4).

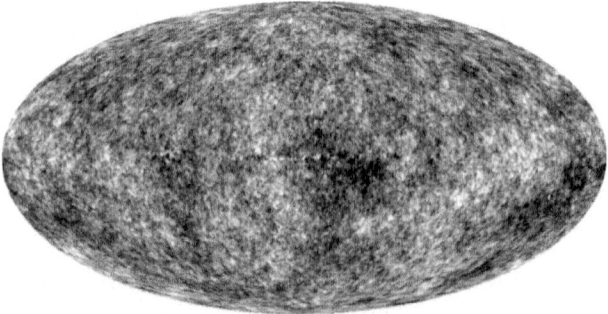

Fig 8.4 (above) -The Cosmic Microwave Background Radiation (CMBR or CMB)

In Fig 8.4, the white areas are high energy density areas, where there are galaxies and their clusters today and the black areas are low energy density areas, where there are 'empty spaces' today. And this 'map' of the universe will remain the same at every point of time.

One may stop here and ask that if the initial universe was very hot, so why does the energy of CMB is below the visible frequency range (means why does the energy of CMB is low)? This is because the CMB radiation is travelling for billions of years (in an expanding universe). As the universe expands, its density decreases. As a result, the energy or mass per unit volume also decreases. Due to this, the temperature (or heat energy) of the universe also decreases. This is why a light wave travelling in vacuum for a long time will lose some of its energy. So, the frequency of the wave will also decrease.

So, this is why the CMB was obtained in radio frequency. The temperature of the obtained CMB radiation is 3 kelvin. The temperature of the universe (in vacuum) is also the same, that is, 3 kelvin. This is the lowest temperature anywhere in the universe except for the 'Boomerang Nebula', which has a temperature of about 1 kelvin.

8.3 The Origin of Our Universe

When Einstein came to know that the universe is expanding, he claimed the cosmological constant was his 'biggest blunder'. But now we know that there is a need for the cosmological constant but its value has been altered. The universe is being expanded by the cosmological constant, that is, the anti-gravity force.

The cosmological constant was not considered physically real, it was considered a mathematical stuff. It was in 1998 when Physicists discovered what is called the '**Dark Energy**' which is now considered the physical reality of the cosmological constant. Dark matter and dark energy (and how they were discovered) will be soon discussed in this chapter.

So, now we shall discuss how the universe originated. The universe is currently expanding, so, if we reverse this process, that is, if we go back in time, the universe would be contracting, and at some point, it would collapse to a single point of zero size, (and hence, infinite density). Such a point of infinite density is called a '**Singularity**'. So, our universe has started from a point, a singularity which we call the '**Big Bang**' and this theory of creation of the universe is called the '**Big Bang Theory**'.

But in this method, one can argue that the universe may have existed forever and at some point of time, something made the universe expand. So, we shall understand it with another method which Stephen Hawking and Sir Roger Penrose used in 1964 to prove that the universe may have started from the Big Bang singularity. Any object will have its past light cone and its future light cone, that is, (absolute) past and future. The past light cone represents the paths of all light waves reaching the object from its past (Fig 8.5).

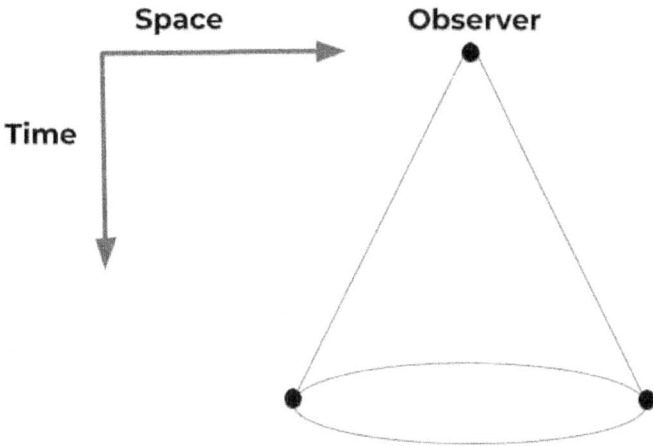

Fig 8.5 (above) -Past light cone of an observer.

The past light cone of an observer represents its absolute past, the past which can affect it in some way.

Now, in the past, the universe would be smaller than it is in the present time. Being smaller means being more dense. When we go further back in time, the density of the universe will be enough to curve space-time and when space-time will be curved, the paths of light will be bent towards each other because everything, including light, follow geodesics in space-time, and ultimately, the light rays meet at a point, the Big Bang singularity (Fig 8.6).

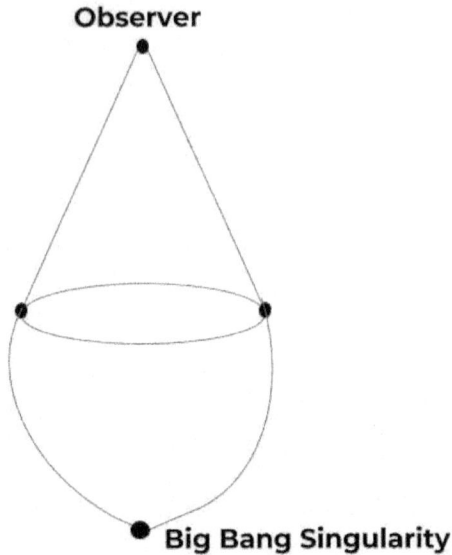

Fig 8.6 (above) -Our Past is pear shaped.

This is how Hawking and Penrose proved, using the general theory of relativity, that our universe may have started from a singularity. There was 'inflation' (not explosion) in the singularity which suddenly created the entire universe.

We now know that the universe started at the Big Bang. But, what actually happened at the time of the Big Bang, that is, at time $t = 0$?

A convincing answer to this can be given by the concept of '**Negative Energy**'. When the big bang took place, equal amounts of positive and negative energy were generated, so the total energy of the universe is zero. So, everything is perfectly balanced, as all things should be!

But, what negative energy actually is? This can be understood with an example. Suppose you are making a small "mountain" by digging the ground. Now, to make the mountain you must have to create a "depression" (hole) in the ground. If we consider that mountain as the positive energy of the universe, the depression will represent the negative energy. But, where is this negative energy now? According to quantum field theory, particles with negative mass can exist for a short period of time (we shall discuss more about this in the next chapter). And, we know that mass and energy are the same thing, so the negative energy may be in the form of these particles. Actually, the gravitational potential energy of any object in a gravitational field is also negative. The gravitational energy of an object in a gravitational field is negative of the product of the object's mass, strength of the gravitational field and the distance between the object and the source of the field.

Gravitational Energy = − (Mass of the object) × (Strength of Gravitational Field) × (Distance of the object from the source of gravitational field)

All these sources of negative mass and energy cancel out and (almost) balance the total positive mass and energy of the universe, so actually the total energy of the universe is, in fact zero (we'll discuss more stuff about this very soon).

Another convincing but somewhat absurd answer is that before the Big Bang, there were no laws of physics. So, there was also no law of conservation of mass and energy. So, anything could come into existence (and in any form) in this case!

8.3.1 The Initial Universe

The universe started with the inflation of the Big Bang singularity.

But, what happened just after the Big Bang? What was the state of the initial universe? How all the things we see today were created?

About fourteen billion years ago (13.8 billion years), the universe, space, time and energy just came into existence and the universe just started to expand, rapidly, in what we call the Big Bang as described previously.

For 10^{-32} **seconds** (about one trillion-trillionths of a second) after the beginning, the universe expanded with faster-than-light speed. This may violate the special relativity, but it can be explained as follows:

"It is actually the space (three-dimensional structure) which expanded faster than light. We know that space doesn't have any mass. Anything which doesn't have a rest mass is allowed to travel faster than light in special relativity. So, space can expand faster than light."

The space is actually still expanding faster than light. Actually, there is a possibility that not only space but time is also expanding with it. This concept will be discussed in more detail in the next chapter.

This rapid expansion which lasted for the first 10^{-32} **seconds** is called '**Cosmic Inflation**' and this idea was suggested by Alan Guth, an American Physicist, in 1979. This inflationary theory predicts the existence of multiple universes 'enclosed' in an 'empty space' called the '**Multiverse**'. The multiple universe theory or multiverse theory is also supported by Quantum Mechanics, according to which there exists every possible universe for every possible set of 'situations'. We'll not go in more detail of this theory here.

Speaking about our universe, the (initial) time period from:

$$t = 0 \ (at \ Big \ Bang)$$

to:

$t = 10^{-43} \ seconds$ (one ten-million-trillion-trillion-trillionths of a second) is called the '**Planck Era**'. The Planck era ended before the ending of cosmic inflation. When the Planck era ended, the universe grew 10^{-35} **metres** (one hundred billion trillion-trillionths of a metre) across. The Planck era is named after 'Max Planck'.

Initially, the universe was so hot that all the fundamental forces of nature (except gravity) were unified, that is, there was no difference in these forces. After some time, the universe expanded, the temperature fell below, so the forces split into **'Electroweak'** force and strong nuclear force. Electroweak force further split into electromagnetic force and weak nuclear force. These are the four fundamental forces of nature (strong, electromagnetic, weak force and gravity) which shape the whole universe.

Strong force binds the atomic nuclei, electromagnetic force holds the nucleus with electron(s), the weak force is responsible for radioactive decay and gravity is responsible for binding the large scale structures of the universe like galaxies and star systems. When all the forces split into the current four individual forces, a trillionth of a second had passed since the beginning.

There was only energy in the very beginning of the universe, but, as energy and mass are the same thing and are interconvertible, so energy in the form of photons started to convert into matter in the form of elementary particles, like electrons, quarks, neutrinos etc.

So, there were bosons (the force carrying particles) and fermions (the particles of matter) in the initial universe. But there were also the antiparticles (described in the previous chapter) for every particle.

The photons at very high energies can convert into a particle-antiparticle pair, like electron-positron (anti-electron), quark-antiquark and neutrino-antineutrino pair. When they come into physical contact they annihilate each other in the form of photons again.

As the initial universe was very 'energetic', photons converted into particle-antiparticle pairs and these particles and antiparticles annihilated each other to form photons, so the universe was 'balanced' as all things should be, but as the universe expanded further, there was not sufficient energy for the photons to convert into particle-antiparticle pairs. So, most of the matter annihilated antimatter, but because for every one billion particles of antimatter, there were one billion plus one particle of matter, so, finally, matter 'won'.

The temperature of the universe was now a trillion kelvin and a millionth of a second had passed since the beginning.

As the universe further expanded, quarks didn't have sufficient energy to escape the strong nuclear force, and combined together due to the strong force to form '**Hadrons**' (described in the previous chapter).

There are two types of hadrons, firstly the baryons which are made of three quarks and secondly the mesons made of one quark and an antiquark.

The baryons mainly consist of protons and neutrons which are the constituents of atomic nuclei. Protons are made of one down quark and two up quarks, and neutrons are made of one up quark and two down quarks, and all the three quarks have different 'colours' — red, green and blue, so protons and neutrons are 'white'.

So now, protons and neutrons were formed. The battle of matter and antimatter was still in play, but it was more like the battle of hadrons and anti-hadrons, and electrons and positrons. Because the number of particles of matter (protons, neutrons and electrons, generally) exceeded the number of particles of antimatter, so matter won.

Now, there was (almost) no antiparticle left which was created in the very beginning by the photons. At this point, one second had passed since the beginning.

Now, the universe had grown a few light years across, about the distance from the Sun to its neighbouring star system, Alpha Centauri, and the temperature of the universe had now fallen to about a billion kelvin.

Electrons still had sufficient energy to escape electromagnetic force. Universe was full of photons. As the universe continued to expand and its temperature dropped to about a hundred million kelvin, protons and neutrons didn't have sufficient energy to escape strong force and protons combined with other protons and neutrons as well, forming atomic nuclei (not complete atoms). Ninety percent of those nuclei were of hydrogen (nucleus of hydrogen is simply a single proton) and the ten percent consisted the nuclei of helium (two protons and two neutrons), deuterium (heavy hydrogen, consisting of a proton and a neutron), tritium (further heavy hydrogen, consisting of a proton and two neutrons) and lithium (a metal, consisting three protons and four neutrons). At this point, about two to three minutes had passed since the beginning.

The electrons still had sufficient energy to escape the electromagnetic force of atomic nuclei. It took **380,000 years** (from the beginning) for the electrons to lose sufficient energy, (when the temperature of the universe dropped to 3000 kelvin), to combine with atomic nuclei to form complete atoms. Electromagnetic force was responsible for this.

After **380,000 years** (from the beginning) the light of the cosmic microwave background (detected by Penzias and Wilson in 1964) was also emitted, which is the earliest visible (observable, more precisely) light.

The temperature of this cosmic microwave background (CMB) is about **3** kelvin (**2.725 kelvin**, precisely).

All these events in the initial universe can be understood by the timeline in Fig 8.7.

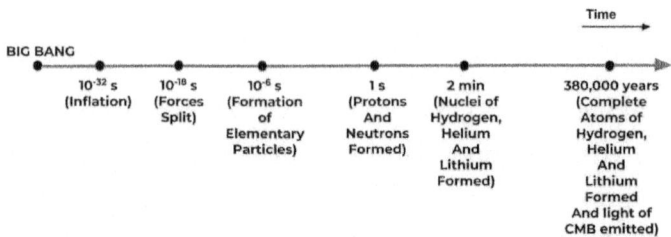

Fig 8.7 (above) - Timeline of Initial Universe (Only Time is represented)

8.4 Evolution of Our Universe

After the formation of the 'complete' atoms of hydrogen, helium and lithium, the hydrogen atoms started to 'bond' with each other, forming molecules of hydrogen gas. Molecules are the structures in which electrons orbit more than one nucleus.

Atoms form molecules to make themselves 'stable'. Helium is already stable, so it didn't create molecules. Helium, in its natural form, is regarded as 'Helium gas'.

Lithium is a metal, metals don't form molecules, instead they form metallic bonds with each other and ionic bonds with other atoms (which contains one atom of a metal and other of a non-metal bound by electromagnetic force).

The universe was full of hydrogen and helium gases and the universe was quite dark because there were no stars, there were huge clouds of hydrogen and helium. This time period when the universe went through a 'dark phase' is called '**Dark Ages**'. Dark ages started **380,000 years** (after the beginning) and lasted for **400 million years** (after the beginning).

After **400 million years** from the beginning, the first stars were born. Stars are created when a huge cloud of hydrogen and helium collapse into a dense sphere due to gravity. Because of being dense, the temperature of this sphere increases and when the temperature reaches a certain limit, nuclear fusion starts in the core of this sphere and the sphere is able to do nuclear fusion of hydrogen to form helium.

When two nuclei combine to form a heavier nucleus, the process is called nuclear fusion. It is actually the opposite of the nuclear fission process (described in chapter 3). This process releases a very large amount of energy, even larger than nuclear fission. The '**Hydrogen Bomb**' or '**Thermonuclear Bomb**' works on this principle. The cause of release of energy is the same as that of nuclear fission, that is, the conversion of mass into energy.

The energy released from nuclear fusion balances the self-gravity of the sphere and the sphere stops to collapse, and this is what we call a '**Star**'. The first stars were formed 400 million years after the beginning and ended the dark ages (all the heavier elements than hydrogen, like carbon, iron, copper, gold, silver etc. are manufactured in stars by the process of nuclear fusion).

The universe was still dark, because the stars were very far from each other; there wasn't any galaxy. The universe was expanding. Some billion years after the beginning, galaxies were formed due to gravity. Gravity of stars attracted them towards each other forming galaxies.

The heaviest object remained at the center of galaxies. This object is a '**Black Hole**', a 'dead star' (we'll discuss more about them soon).

Formation of galaxies ended the dark ages 'completely', giving us the universe we see today. Our Milky Way galaxy was created some one billion years after the beginning, that is, some **13 billion years** ago from today . It is one of the oldest galaxies known.

It was **9 billion years** after the creation of the first galaxies, that an average sized star, **Sun**, was born. Sun was born about **4 billion 600 million years** ago and is expected to live for another 5 billion years.

Some **100 million years** after Sun's creation, a rocky, small sized planet **Earth** was formed (about **4 billion 500 million years** ago). And 300 million years after the Earth's creation, life started (some 3 billion years ago). The first generation of life was very simple (single prokaryotic cells etc.), then life evolved into complex organisms; extinctions and mass extinctions took place.

Finally, **2 million years** ago, **Homo** Sapiens (human beings) came into being.

In this section and the previous section 8.3.1, I have summarized the evolution of the universe from the Big Bang till now. In the next section of this chapter, stars and their life will be discussed.

8.5 Life of Stars

Stars are the most important objects in the universe. They manufacture all the heavy elements which are important for us, like iron, copper, aluminium etc. They manufacture the elements we are made up of, like, nitrogen, carbon, iron (again), phosphorus etc.

When a very big cloud of hydrogen and helium collapses into a (relatively) small sphere, a star is formed. The cloud collapses due to its gravity (of course). When the sphere is hot enough to start the nuclear fusion process in it, but gravity still dominates, the stage is called a '**Protostar**'. The temperature of the collapsing cloud increases because of the increasing density of the gases. Due to increasing density, the particles of the gases come very close to each other. This leads to increased friction which generates heat (actually it's due to the collision of the particles).

When sufficient energy is produced by nuclear fusion to balance gravity, the star becomes stable and consistently carries out the process of nuclear fusion inside its core for years. This stage is called 'Main Sequence Star'. [Nuclear energy balances gravity as nuclear energy is pushing the sphere outwards and gravity is pushing the sphere inwards.]

Main sequence stars can be of two types — **Massive Stars [of Blue and Blue-white Colour(s)]** and second are (comparatively) **Less Massive stars [of Red, Yellow-red, Yellow and Yellow-white (like Sun) colour(s)]**.

8.5.1 Spectral Classification of Stars

Stars can be classified on the basis of their colour, luminosity and heat. This classification is called **'Spectral Classification'**. On the basis of the above three factors, stars are classified into eight different categories – **O, B, A, F, G, K** and **M**.

This order is in decreasing luminosity and heat, that is, 'O' type stars are most luminous and hottest while 'M' type stars are least luminous and coolest.

So, 'O', 'B' and 'A' type stars are the hottest and most luminous. Being most luminous and hottest means that they are most 'energetic'. So, how can we find out the colour of the star using this data? In the spectrum of 'VIBGYOR' (Violet, Indigo, Blue, Green, Yellow, Orange and Red), the light with colour(s) of blue group (Violet, Indigo and Blue) has the highest frequency and hence, the highest energy. So, the colour of 'O', 'B' and 'A' type stars is blue.

'F' type stars are less 'energetic' than 'A' type stars, so, according to VIBGYOR, they have a blue-white or yellow-white colour (not green).

'G', 'K' and 'M' type stars are the least luminous and least hot, so according to VIBGYOR, they have a colour(s) of yellow, orange or red. Our Sun is a 'G' type star.

8.5.2 Stages in the Life of a Star

The first stage in a star's life is a protostar, as described previously. Then comes the main sequence, the longest stage in a star's life. Main sequence star can be of any spectral classification and, the greater the frequency (and energy) of the star's colour, the larger its size and mass.

This means 'O' type main sequence stars are the largest and most massive main sequence stars and

'**M**' type main sequence stars are the smallest and least massive ones.

In the main sequence, there are some 'dwarf' stars which usually belong to the spectral group '**F**', '**G**', '**K**' or '**M**'. The '**Yellow Dwarfs**' (like Sun), belong to '**F**', '**G**' or '**K**' groups. The Sun is a '**G**' type yellow dwarf (but the Sun is yellow-white in colour instead of yellow).

The '**Red Dwarfs**' [like Proxima Centauri (Sun's closest neighbour)], belong to the '**K**' or '**M**' group. Proxima Centauri is an '**M**' type red dwarf.

The blue and blue-white main sequence stars are much larger than red and yellow main sequence stars and live shorter than red and yellow stars. This is because they are large and massive, so they have to produce more amount of nuclear energy to balance gravity. So, they exhaust their 'fuel of nuclear energy' faster to maintain the balance, and thus they 'die'.

Any type of star, after exhausting their hydrogen, starts to burn other elements like helium which produces heavier elements like carbon, oxygen etc. Then they start to burn these heavier to produce further heavier elements. Actually a star doesn't really run out of a particular fuel. For example, when a star is about to die it is still converting hydrogen into helium in the outer part of the core; a little further down, helium into carbon and oxygen; further down, silicon into iron.

When red and yellow dwarfs exhaust their hydrogen, they become about 100 times bigger in volume and are called '**Red giants**'. More bigger red and yellow main sequence stars become '**Red Supergiants**'.

Blue and blue-white big main sequence stars, after exhausting hydrogen become '**Blue Giants**', '**Blue Supergiants**' or red supergiants.

A star dies when iron is produced in its core, because fusion of iron is not really something that stars can do. The star doesn't have enough energy to fuse iron (because, it requires energy to fuse some type of atoms, and, the heavier the atoms, the more energy is required to fuse them). So, the star just can't balance gravity and the core starts to collapse under its own gravity.

When red giants die, the event isn't so violent, the core just 'leaves' other parts of the star including all the elements inside it and spreads all the manufactured elements in the universe in a peaceful manner, and the remaining 'dead body', is called a '**White Dwarf**', white in colour, which have mostly carbon inside it.

If the mass of a white dwarf is smaller than a certain limit, called the '**Chandrasekhar Limit**', then the white dwarf is stable and will live forever as a white dwarf or will convert into a '**Brown Dwarf**'. If the mass of a white dwarf is greater than the Chandrasekhar limit, it explodes in a very violent explosion called a '**Supernova**' and then becomes a '**Black Hole**' or a '**Neutron Star**' in some cases.

(Subrahmanyan Chandrasekhar was an astrophysicist who was awarded the Nobel Prize in Physics in 1983 for his work on the Chandrasekhar limit.)

Chandrasekhar limit is equal to 1.4 times the current mass of the Sun.

Big red supergiants, blue giants and blue supergiants explode in a very violent supernova and then spread the elements manufactured in them across the universe, and leave their 'dead body' as a '**Neutron Star**' or '**Black Hole**'.

Stages in a star's life can be understood by Fig 8.8.

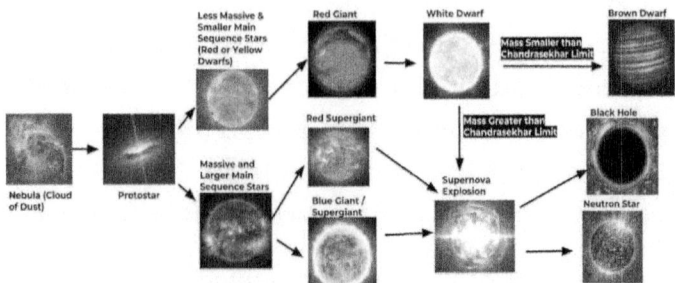

Fig 8.8 (above) - Stages of a Star's Life

Neutron stars are highly dense objects. When a large star starts to collapse under its own gravity, protons get 'merged' with electrons due to high pressure to form neutrons (and neutrinos), that's why it is called a neutron star because it contains almost only neutrons in it.

The neutron stars have a very strong magnetic field. There are two special types of neutron stars — '**Pulsars**' and '**Magnetars**'. Pulsars emit light beams from their poles due to strong magnetic fields. Magnetars have the strongest magnetic field in the universe, even stronger than black holes. Neutron stars are very small in size, smaller than the Sun, but their mass is much larger than that of the Sun, this is due to their density, that is, they are highly dense.

A black hole is a dead star and is often defined as a region in space-time where gravity is so strong that nothing, even particles and light can escape from it.

The idea was started in 1783 by an English philosopher, John Michell. He suggested that there can be a large number of very massive stars, so massive that the light emitted by them is instantly dragged back by the star's gravity, so they are invisible.

When it was proved in the nineteenth century that light is a wave, this idea seemed to be incorrect, because waves don't have mass, a wave is a disturbance, a transfer of energy.

In 1905, when Einstein proved Planck's hypothesis (that light also has a particle nature) by explaining the photoelectric effect, this idea of black hole seemed to be true. But it was only in 1915, after the development of the general theory of relativity, when this idea gained some support.

Actually, in general relativity, gravity is a distortion in space-time. When a region in space-time is curved, the path of a straight moving object (in the curved space-time) appears to be curved, the same phenomenon happens with light.

Light bends due to gravity, more precisely, due to curves in space-time, or in other words, light is affected by gravity. This is also called '**Gravitational Lensing**', because gravity affects light like a lens (Fig 8.9).

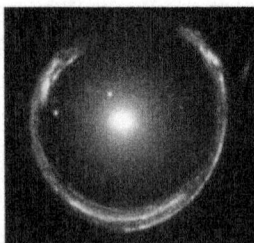

Gravitational Lensing

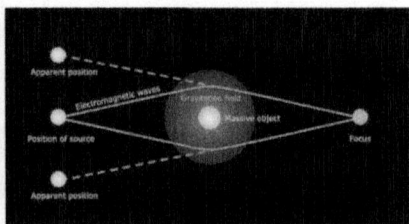

Mechanism of Gravitational Lensing (Light is a form of Electromagnetic Wave)

Fig 8.9 (above) - Gravitational Lensing and its Mechanism

Black holes affect light in a different way, they trap light in them. Black holes curve space-time infinitely. A representation of a black hole is in Fig 8.10

('**Schwarzschild Radius**' is just the distance between singularity and any point on the event horizon).

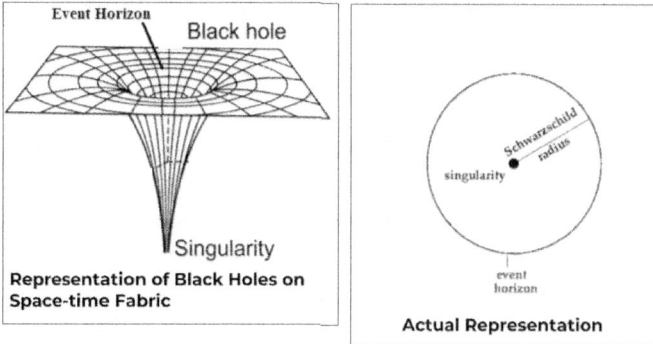

Fig 8.10 (above) - Representation of a Black Hole

Fig 8.10 can be compared with a waterfall. The point where the water just falls is equivalent to '**Event Horizon**' (which is the boundary of black hole), the black part of the black hole (Fig 8.11). If one reaches this point, one can never come back.

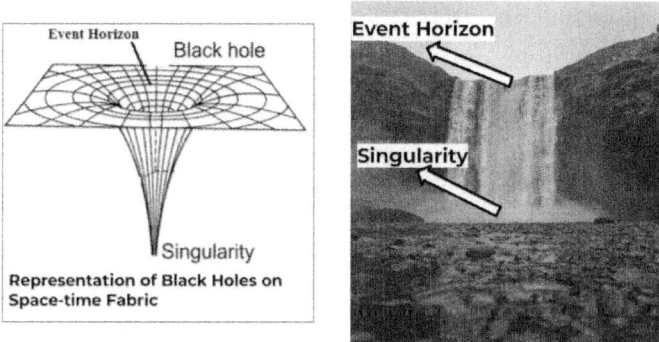

Fig 8.11 (above) - Black Holes and Waterfall

The lowest point in a waterfall, that is, the point where water touches the ground is equivalent to the singularity of the black hole. Singularity, as described previously, is a point of infinite density.

Actually, Schwarzschild radius can be calculated for any object. If an object is shrunk into a sphere of radius equal to its Schwarzschild radius, it becomes a black hole. Mathematically, Schwarzschild radius of any object is twice the product of Newton's Universal Gravitational constant and the mass of the object divided by the square of the speed of light.

Schwarzschild Radius = (2 × Universal Gravitational Constant × Mass of the Object)/(Speed of Light Squared)

This above equation is just an approximation as it is derived from Newton's theory of gravity.

A singularity covered by an event horizon is called a '**Closed Singularity**', for example, a singularity of a black hole, and, the singularities which are 'open' are called '**Naked Singularities**', for example, the Big Bang singularity.

Closed singularities are always in the future of 'anything', that is, closed singularities only have a past light cone and not a future light cone and naked singularities are always in past of 'anything', not in the future, for example, Big Bang singularity is in our past, we can't go back before the Big Bang. Naked singularities only have a future light cone but not a past light cone (Fig 8.12).

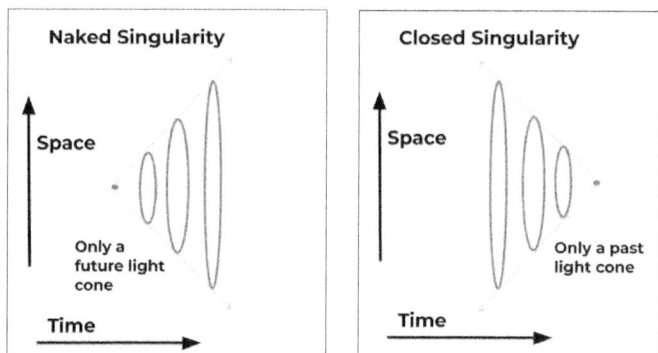

<u>Fig 8.12 (Above)</u>

More about Black holes (and its first 'real image') will be discussed in more detail in the chapter 'Relativity and Time Travel'.

8.6 The Dark Universe

Does the familiar matter constitute the whole universe? Or is there something like 'invisible matter', which we can't observe directly? Yes! There is, and, it is expected that this invisible form of matter and energy constitutes 95% of the whole universe, this means that the matter, antimatter and other forms of observable mass and energy constitutes only 5% of mass and energy in the universe.

These mysterious forms of matter and energy are known as '**Dark Matter**' and '**Dark Energy**'.

The invisible mass is called dark matter. We know that light is bent due to gravity (gravitational lensing). A galaxy is very massive, so, sometimes light is 'wrapped' around it and forms a 'ring'-like-structure called an '**Einstein's Ring**' (Fig 8.13).

A similar Einstein's ring was observed around a galaxy. It was also observed that the galaxy didn't have sufficient mass to create this Einstein's ring. It was suggested that the remaining gravity is provided by a different type of matter — '**Dark Matter**'. This matter doesn't interact with light and normal matter (and antimatter) through the nuclear forces and the electromagnetic force. It only interacts through gravity. It means dark matter has mass. Dark matter has gravitational effects. 23 percent of the mass and energy in the universe is dark matter (I am really using the phrase 'mass and energy' again and again to make you believe that these are the same thing).

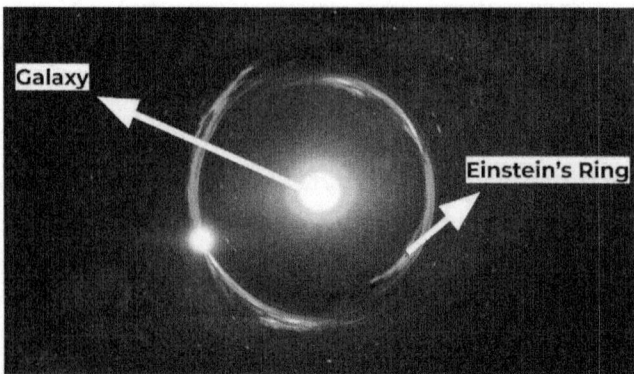

Fig 8.13 (above) - An Einstein's Ring

In 1998, a team of astrophysicists observed several supernovae but they were dimmer than the team expected. More precisely, they got more red-shifted than expected. This was a clue that the universe is accelerating faster than expected. This means, the expansion of the universe (after cosmic inflation) wasn't uniform, but the universe is expanding more rapidly than it did in the dark ages. It was suggested that a mysterious energy, **'Dark Energy'**, is responsible for this.

Actually, Einstein 'threw out' the cosmological constant from his equations of general relativity when Hubble discovered the universe is expanding, but in 1998, it was put back in the equations and with a numerical value of **11×10^{-53}**. Now the cosmological constant had gained a 'physical reality' in the form of dark energy.

Dark energy is not an anti-gravity force, it interacts with gravity like dark matter but, dark energy is accelerating the universe. The reason is still not clear completely as the research on both the dark matter and energy is still going on.

A recent research has tried to unite dark energy and dark matter as **'Dark Fluid'**. It is like an alternative theory to dark matter and dark energy.

Theory of dark fluid proposes that dark matter and dark energy are not separate physical phenomena as previously thought, nor do they have separate origins, but they are strongly linked together and can be considered as two facets of a single fluid. At galactic scales (or structures smaller than galactic scales), the dark fluid behaves like dark matter, and at even larger scales its behavior becomes similar to dark energy.

This dark fluid, currently, is only a hypothesis and this is being developed currently. The first paper describing dark fluid was published in December 2018.

8.7 Friedmann's Models of the Universe

Using his equations (described previously), Alexander Friedmann, gave three possible models of the universe.

These describe the possible 'shapes' of a universe. The shape depends on '**Density Parameter'**. Density parameter is equal to the current amount of mass-energy in the universe divided by the amount of mass-energy required to stop the expansion of the universe. The shape of the universe can be flat (**no curvature and zero energy**), spherical (**positive curvature and positive energy**) or something like a saddle (**negative curvature and negative energy**). Here the 'curvature' refers to the overall curvature of space-time and the energy refers to the total energy of the universe.

But these shapes are not meant in a 'literal' sense. The space-time is four dimensional and thus, we can't perfectly represent a curvature in it. But, we have different things and analogies to understand this all.

A **'Flat Universe'** (having no overall 'curvature') means that if we imagine a triangle spanning between the edges of the observable universe, the sum of all angles of that triangle would be **180°**, which is only true in a flat space. This is because if you will draw a 'perfect triangle' on a flat sheet of paper, the sum of its angle will be **180°**. This is also known as the **'angle sum property of a triangle'** (this is an elementary thing), according to which the sum of all the interior angles of a triangle is exactly **180°**. Fig 8.16 illustrates a flat universe.

Similarly, a **'Spherical Universe'** (having 'positive curvature') means that if we imagine a triangle spanning between the edges of the observable universe, the sum of all the angles of that triangle would be between **180°** and **270°**. For example, if you draw a triangle on a perfectly spherical ball, the sum of all its interior angles will be **270°**. Fig 8.15 illustrates a spherical universe.

Unlike a flat and spherical universe, a universe with a shape of saddle is difficult to imagine due to a 'negative curvature'. In such a universe, the sum of angles of a triangle spanning between the edges of the universe will be less than **180°**. Fig 8.14 illustrates a universe taking the shape of a saddle.

Now, we should discuss how the values of the density parameter affects the shape of the universe and its possible end.

If the amount of mass-energy required to stop the expansion of the universe is greater than the amount of mass-energy presented in the universe currently (density parameter is smaller than one), then the universe will take the shape of a saddle and will expand forever with a very high rate of expansion. The total mass and energy of this kind of universe is negative.

In this case the universe will face a rapid expansion and all the galaxies will 'break down' at a certain point of time. The universe will become 'inactive' or 'dead'. Only space and time will exist, but there will be no meaning of them.

When the universe 'dies' or 'ends' in this way, it is called a '**Big Rip**'. Fig 8.14 illustrates what will happen if the density parameter is smaller than one.

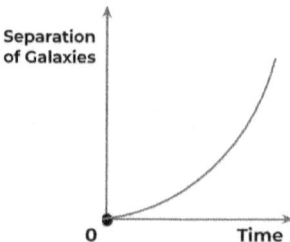

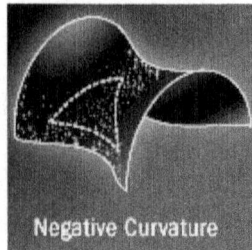

Fig 8.14 (above) - Density Parameter smaller than one, a 'saddle-shaped' universe.

If the amount of mass and energy required to stop the expansion of the universe is smaller than the amount of mass and energy presented in the universe currently (density parameter is greater than one), then the universe is spherical and the speed of expansion of the universe will start to decrease, at a certain point in time, the universe will stop expanding and will start to collapse into itself, finally ending as a singularity called the '**Big Crunch**' singularity. Fig 8.15 illustrates what will happen if the density parameter is greater than one. The total mass & energy of this kind of universe is positive.

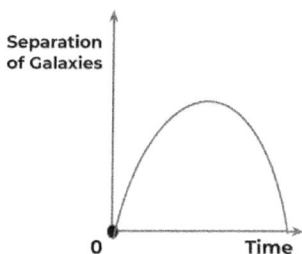

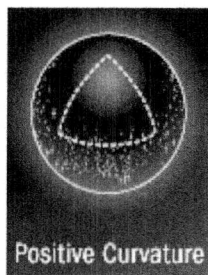

Fig 8.15 (above) - Density Parameter smaller greater one, a 'spherical' universe.

If the amount of mass and energy required to stop the expansion of the universe is equal to the amount of mass and energy presented in the universe currently (density parameter is equal to one), then the universe is flat and will remain flat forever. Moreover, the total mass & energy of this kind of universe is zero.

It will expand forever but not as violently as in a Big Rip. The universe will expand forever like in a Big Rip and its end will be similar as Big Rip but it will last longer than that. The universe, in this case, will end as a '**Big Freeze**' (also called '**Heat Death**'). Fig 8.16 illustrates what will happen if the density parameter is equal one.

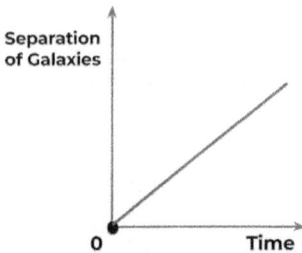

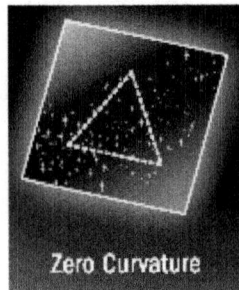

Fig 8.16 (above) - Density Parameter equals to one, a 'flat' universe.

Now, the question is what the shape of our own universe is and what is its end? When we add all the (observable) mass and energy, and find out the value of the density parameter, it remains smaller than one. When we add dark matter, it still remains smaller than one. But, as dark energy also has gravitational effects, it should also be added. When the amount of dark energy is also added, we get the value of the density parameter (almost) equal to one.

This means that our universe (not earth) is flat. In a flat universe, the total energy is zero, that is, the total positive mass and energy of our universe is equal to its total negative mass and energy.

So, our universe will end in a Big Freeze. The Big Freeze is supposed to happen after some 10^{106} **years** from now.

The next chapter is based on the topic of time travel (I wasn't really going to include this stuff in this book, but this is a trending topic among general readers, so I just decided to do that!)

9

RELATIVITY AND TIME TRAVEL

9.1 Introduction

We have discussed a number of things now in this book. We have seen how our understanding of the universe was revolutionized in the twentieth century by Relativity and Quantum Mechanics.

We haven't discussed '**Time Travel**' yet. Time Travel is a very trending topic for science fiction movies, but, actually, according to various principles of Physics, it is possible.

We shall discuss the possible ways for time travel, according to Special Relativity (SR) and General Relativity (GR) (there will also be some "how-to time travel guides" so that you can do it yourself and fix your past or see how you'll be doing in future). We shall also discuss what time actually is and the possible reason for the flow of time.

9.2 What is Time and Why does it Flow?

We have used the word 'time' more than 200 times in this book. We all are quite familiar with the measurement of time. But, what is time? Why does it flow? And, why does it go forward, not backward?

As we discussed in Chapter two, time is a dimension. But, why is it not like dimensions of space?

For example, we travel in space dimensions every day. When you are in motion, you are travelling in space. Travelling in space dimensions means covering some 'distance'. It is very simple. But, it is not as simple to travel in time. For example, if you want to go backwards in time, you can't just turn back and start moving!

So, that's why time is a different type of dimension. Why is it like this? Well, nobody probably knows!

Also as time moves on, it leads to development or destruction (we can refer destruction to as negative development).

The other question is why does time flow? Moreover, why does it flow in a forward direction?

An excellent answer for the flow of time was given by Professor Richard A. Muller (Physicist) in 2016, in his research paper titled '**Now, and the Flow of Time**'.

The paper was co-written by Shaun Maguire (Physicist).

Muller proposed that the flow of time can be easily understood by assuming the expansion of the universe in all the four dimensions of space-time rather than in only three dimensions of space. This means that the universe is also expanding in time.

What does it mean now? The paper proposed that space and time are so strongly linked in relativity that when new space is created, new time is also created.

We know that the universe is expanding. It means that the three-dimensional space is expanding. The space between galaxies is expanding. This means that new space is being created every moment between the galaxies. So, with the creation of new space, new time must be created. This is why time flows. "New time is created every moment". Now, at this moment also, some new time is created!

So, if the universe weren't really expanding, time would not be passing or flowing. Every single object would simply stop as we see in science fiction movies that there is a person with a clock. He presses a button on the clock and everything stops because of stopping time. Same would be the case if the universe were static.

We can take another example to prove Muller's theory. If the universe were contracting, it would finally reach the point of zero size, that is, a singularity. We also know that the universe began at a singularity. Also, according to Muller's theory, time would reverse if the universe were contracting and the universe would finally reach the point where it began (at time, $t = 0$), that is, a singularity! So, we can say if the Big Bang theory and general relativity are correct, then Muller's theory may also be correct.

So, now we have probably got the answer to the questions raised in this section. Now, we can move on to time travel.

9.3 Time Travel in Special Relativity

We discussed the concept of 'time dilation' in Chapter three, 'The Relativistic Effects'. This principle of 'time dilation' can be used for time travel.

According to special relativity, time slows down for a moving observer. This effect is observable at very high velocities — velocities near to that of light. Travelling at such high speeds can take us to the future!

Suppose there is a road along the equator of the Earth. Now, there is a car which can travel up to ninety eight percent of the speed of light. Now, a person gets inside the car and he just starts to drive the car in the year of 2020 at the maximum speed. He doesn't face any obstacles (he may have faced some, as he started in the year 2020); he just travels at a constant speed (ninety eight percent of the speed of light). He travels for this like one year (one year, according to him).

For him, the year should be 2021, but actually he realizes that the year is something more than 2021 (one can calculate this using the equation of time dilation). It means he has travelled to the future! This is because of the fact that time has slowed down for him as compared to other people on Earth. (Now, you may need to stop here, go back to chapter 3 and read the time dilaton thing again to understand what this really means, because people often forget the real meaning of time dilation.)

Particles are time travellers, they can travel near the speed of light. So, time slows down for them (and in the case of photons, time stops down).

But, can one travel to the past using the same principle? The answer is yes if one can travel at a speed faster than light. But, we know that special relativity doesn't allow us to do this as it would require an infinite amount of energy. But, for a second, just assume that it's possible to travel at the speed of light and even greater than it.

Now, assume that a person is travelling with the speed of light. At this moment, time wouldn't be passing for him. Now, as he exceeds the speed of light, time will start to run backwards for him (but this can't happen in reality as anything having some rest mass can't travel or exceed the speed of light). But, how have we concluded that time will stop for him or time will run backwards for him?

The answer is simple. This is just simple logic. As the velocity of an object increases, time dilation increases for the object. That is, time passes more slowly. As it reaches the speed of light, time is dilated infinitely, that is, it stops. If it exceeds the speed of light, time will start to run backwards, that is, in the negative direction.

9.4 Time Travel in General Relativity

In the general theory of relativity, there are two ways of time travel. We shall firstly discuss the method of **'Gravitational Time Dilation'** and then **'Wormholes'**.

9.4.1 Gravitational Time Dilation and Time Travel

We discussed the concept of gravitational time dilation in chapter six, 'Curved Space-time and Geodesics'. Any source of mass and energy dilates time, that is, near a source of mass or energy, time is dilated, and this dilation decreases as we go away from that source. Also more the mass or energy, the greater this effect.

So, we can travel to the future by coming in contact with a strong gravitational field. For example, suppose a person from earth goes near a neutron star in a spaceship. He starts to orbit the neutron star at a safe distance. Now, as he will come back on earth, after orbiting the star for some years, he will find that the total time he took to go to the neutron star, the time for which he orbited the neutron star and the time he took to go back is less than that passed for the people living on the earth.

So, if the space traveller left the earth in 2020, took one year to reach the neutron star, orbited the neutron star for one year and took the same time (one year) to go back to the earth (note that these time periods are measured by that space traveller), he will realize that on earth the year will not be 2023 (as it should be according to the person), the year would be 2024 or 2025 or later, depending on the strength of gravity of the neutron star.

Note that the neutron star can be replaced by any other source of strong gravitational field.

9.4.2 Wormholes and Time Travel

In this section, we shall discuss what wormholes are and how they open a possibility of time travel.

Before discussing wormholes, we should discuss 'Black Holes' and 'White Holes'.

We have already discussed something about black holes in the previous chapter, including their history, the Schwarzchild radius and what they basically are.

If we talk about the components of black holes, they broadly have two components —

1. **Event Horizon** (the boundary of black hole), which is the black part of the black hole. If we see a black hole from outside, we only see its black part, the event horizon.

2. **Singularity**, which is at the centre of the black hole. It is actually the dead star's core which is compressed into a zero sized object (due to which it has infinite density, and gives an infinite curvature to space-time).

The distance between the singularity and any point on the event horizon of a black hole is known as its **'Schwarzschild Radius'**, named after the German Theoretical Physicist, 'Karl Schwarzschild', who had a great contribution in the development of our understanding of Black Holes.

A quite interesting fact about black holes is that neither they are 'holes' nor 'completely black'! Yes, they are not completely black actually. The reason behind this is that black holes have a very strong gravity. So, any cosmic dust around the black hole is attracted towards it.

It starts to orbit the black hole due to gravity. But, it doesn't fall instantly into the black hole, due to a 'pseudo-force' known as the '**Centrifugal Force**'.

Centrifugal force arises when anything is in 'rotational motion' that is, rotating or orbiting something. It "forces" the rotating object 'outwards'. It is actually an inertial force just like the force felt by a person inside a car when the car suddenly stops or starts moving. Centrifugal force 'arises' due to the constant change in direction of the rotating object (due to inertia of direction basically). It's actually not a 'real' force. Centrifugal force can be easily understood by Fig 9.1.

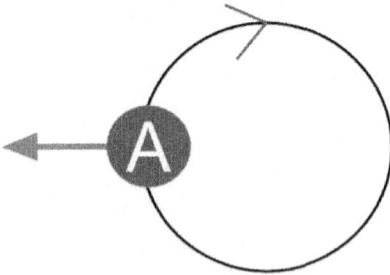

Fig 9.1 (above) - Centrifugal force. In the figure, the circle represents the path of object 'A' and the red line shows the direction of centrifugal force.

The same case is with the dust orbiting the black hole. It should be noted that after a certain period of time, the dust will fall inside the black hole. And, this case is the same with all objects orbiting the black hole.

For example if a black hole is just to 'eat' a star near it, the star will firstly rotate for sometime around the black hole and then will fall inside. But, it will not take much time because the star is already very massive and this creates a strong gravity between the black hole and the star, and this gravity will quickly overcome the centrifugal force.

But, how black holes aren't completely black? Now, as we may know, the dust and gas rotates around the black hole for a very long period of time, and its speed of rotation will be very fast. Due to this, the gas and dust will develop a high amount of frictional force because the dust particles will collide with each other.

We know that friction leads to the generation of heat. The heat energy produced in the dust converts into light energy (like the heat energy converts into light in a light bulb), due to which it starts to glow with quite high intensity and brightness. This is why black holes aren't black due to the bright disk (this disk is known as '**Accretion Disk**') of dust revolving around it.

This is illustrated in Fig 9.2, which is the first 'real image' of a Black Hole captured by the '**Event Horizon Telescope**'. Katie Bouman, an American computer scientist, has a major contribution in the development of the image of this black hole. She led the development of an algorithm for imaging black holes.

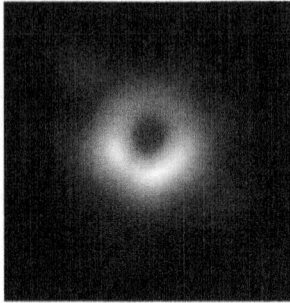

Fig 9.2 (above) - The first real image of a black hole developed by Event Horizon Telescope. The Event Horizon (the black part) and the Accretion Disk can be clearly seen in the image.

The black hole in Fig 9.2 is a '**Supermassive Black Hole**' located at the center of **M87 (Messier 87) Galaxy** and is about **7 billion** times as massive as the Sun. Somewhat similar black hole by the name of '**Sagittarius A Star**' lies at the center of our galaxy, the Milky Way, and is about 4 million times as massive as the Sun.

For years, it was believed that black holes exist forever. Of course, black hole is already a dead star, so, how can it die! However, in 1974, theoretical physicist, Stephen Hawking provided strong theoretical arguments for the fact that black holes 'die', more precisely, they 'evaporate'.

The mechanism behind this is very easy. Actually, the 'empty vacuum' is not really empty. There are quantum fluctuations in the so-called empty space. It means that some virtual particles appear and then suddenly disappear (mostly a pair of particle and anti-particle appear and then they both annihilate themselves).

So, think about this, a particle appears near a black hole. Now, the black hole 'eats' the particle. And, according to the quantum field theory, particles with negative mass can exist for a short period of time. Now, suppose the particle our black hole ate was negative in mass.

Suppose that the particle had a mass of negative of **9 × 10⁻³¹ kilograms**. Now as it was negative, the black hole will lose **9 × 10⁻³¹ kilograms** of its mass.

It loses it in the form of energy (or radiation) coming out of it. This is a very small amount of mass, that is why, black holes die in billions of years. This process is called '**Black Hole Evaporation**' and the energy (or mass, both are the same thing) it loses through radiation is called the '**Hawking Radiation**' after the name of its discoverer.

Hawking Radiation and Black Hole Evaporation are illustrated in Fig 9.3.

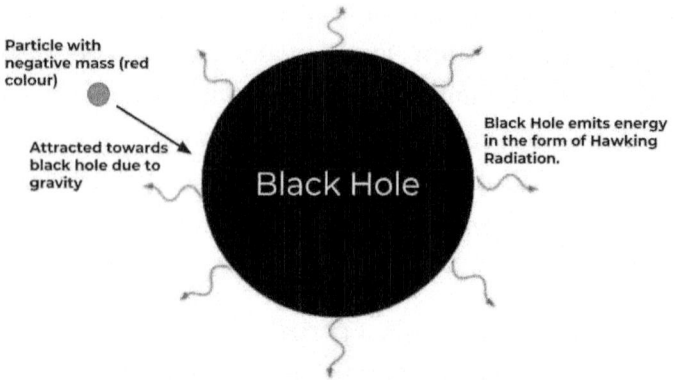

Fig 9.3 (above) - Hawking Radiation (Black Hole Evaporation)

The explanation of black hole evaporation we just discussed above has a little problem with it. The mass of a black hole is positive (obviously) and the mass of the particle (in consideration) is negative, so, by principle, the black hole should repel the particle instead of attracting it. But, by how much force? It should repel this particle with the equal force by which it would attract a particle of the same positive mass. But, why should it repel it? Because, in gravity, opposites repel and likes attract. So, the particle must not pass through the event horizon.

We also have another explanation of this phenomenon, which might seem more correct. According to quantum field theory, a pair of any virtual particle and its anti-particle can appear randomly in space.

So suppose, such a pair appears near a black hole. One of the particles from the pair is 'eaten' by the black hole and the other, somehow, escapes it. Now, to prevent the violation of the conservation of energy (or mass), the black hole must lose some mass. And it loses it in the form of 'Hawking Radiation'.

There is yet another explanation for this phenomena, according to which the black hole 'converts' a 'virtual' particle into a 'real' particle due to which the black hole loses energy. We may not get further into this idea (as you may know, we don't have the 'space' to discuss this more).

9.4.2.1 A Trip to Black Hole!

Going inside a black hole for anyone (or anything) may not be a good experience!

But just suppose, a person decides to take a trip to a black hole (maybe he has been exhausted with his life or something). This is not completely clear even today what will be the experience of going inside a black hole and reaching the singularity.

As described in the chapter, 'The Uncertainty Principle and Quantum Physics' that at a singularity, the physics we currently have fails, because singularity is the meeting point of a large mass (infinite density, which should be described by GR) and very small scale (infinitely small, which should be described by QM). But, as described previously, when GR and QM are applied at the same time, very strange results are obtained like the universe should have collapsed instantly now (but the universe is 'alive')!

So, now, suppose, he is going near to it. As described previously in this chapter, as he will go near the black hole, time will start to dilate for him because black hole has an extremely strong gravitational field.

Now, as he will go more near, time will dilate more and more. He himself will observe no change. This effect will be observed by any observer far from the black hole at a safe distance.

Now, as described previously, he will start to orbit the black hole and will avoid going through the event horizon because of the strong centrifugal force, so strong, that he will die! But, for a minute, think that he has got the guts to survive this much strong force and can observe everything around him.

As he reaches the event horizon, time will stop for him! The observer observing him will observe that the person is not moving and he is stationary at the event horizon.

But, he will actually go through the event horizon, inside the black hole. The event horizon is actually referred to as 'the point of no return'. This is because of the fact that even light can't escape a black hole, and the speed of light is the cosmic speed limit. So, nothing can escape a black hole.

Now, as he has crossed the event horizon, he will die, because of '**Spaghettification**', a process in which an object is stretched due to difference in gravity applied on top and bottom point of the object (Fig 9.4).

He will be spaghettified because there will be a very huge difference in the gravity applied on his legs and on his head by the black hole, which will stretch him, and he will instantly become a part of the singularity!

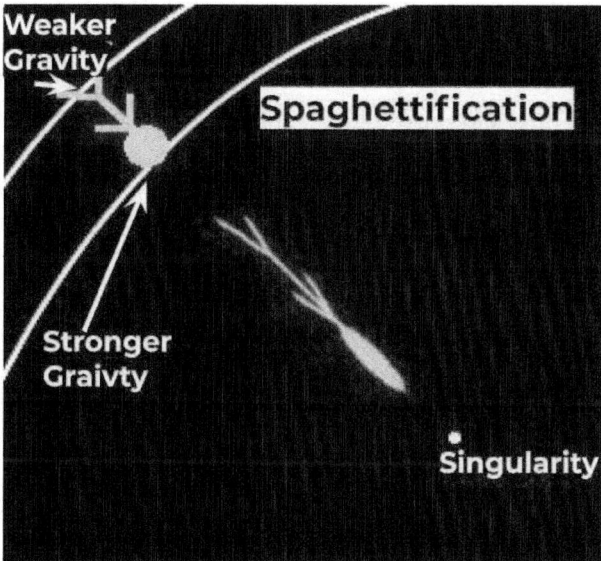

Fig 9.4 (above) - Spaghettification

It wasn't a quite good experience for him. It may be even worse, because some theories suggest that the event horizon of a black hole is covered by a fireball, literally a ball of fire. So, the person will be instantly burnt.

9.4.2.2 Einstein-Rosen Bridge

Now, we can come on to the main topic of this section, **'Wormholes'** or **'Einstein-Rosen Bridges'**, named after their discoverers, Albert Einstein and Nathan Rosen.

Before arriving on this topic, we shall first discuss a brief topic of **'White Holes'**. Actually, white holes are just opposite of black holes. There is a possibility that the singularity of a black hole may be linked with the singularity of a white hole. In this way, when a black hole eats any object, the object will be 'thrown out' by a white hole.

Unlike a black hole, a white hole gives a 'negative curvature' to space-time (Fig 9.5).

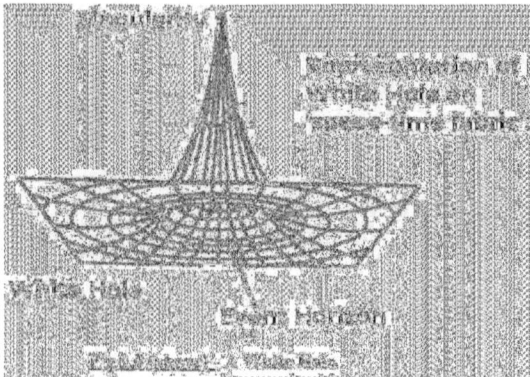

A wormhole is generally referred to as a 'portal' in science fiction. At one end of a wormhole, a black hole is present and at the other end, a white hole. This is how a wormhole works.

So, wormholes are 'bridges' in space-time. They link two points in space and time. For example, a wormhole may link to a point in the past or future. So, now the question is why we don't simply open a wormhole to perform time travel.

Now, as you may have seen in science fiction, energy is needed to keep a wormhole open. Actually, a very large amount of energy is required to do this, which we can't do.

But, really a number of wormholes are present in the universe, even around you! But, they are very small. They are present on a quantum scale. Only very small particles like '**Higgs Boson**' can travel through these quantum wormholes.

A wormhole is represented in Fig 9.6.

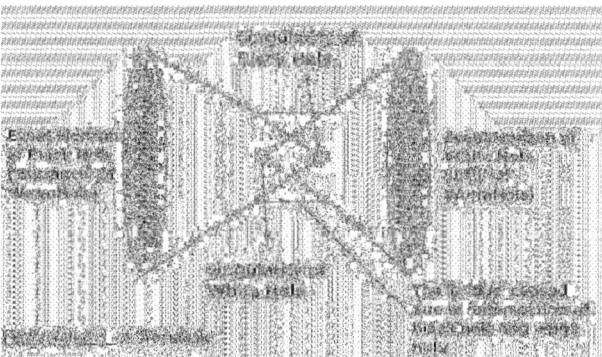

9.5 Paradoxes in Time Travel in Past

We have now probably discussed almost all the possible ways of time travel both in the future and past, but some paradoxes or contradictions arise in time travel to the past.

For example, consider a person who goes to the past (through a wormhole), and meets his great grandfather. He kills him by shooting him with a gun! Now, as his ancestor is killed, he should also not exist. So, his existence is erased. The paradox in this is that if he doesn't exist, who shot his great grandfather? This is a popular one known as the '**Grandfather Paradox**' (another paradox is the **Bootstrap Paradox**, in which somebody gives you a random object, you go in the past and give it back to them, now the origin of this random object is unknown).

Another argument against past time travel is that, if time travel to the past is possible why have the time travellers from the future not come in our time yet?

Now, these are some reasons that time travel may not be possible to the past. But, time travel to the future is probably possible.

Now, we have ended all the chapters of this book and we shall now head to the '**Conclusion**' and '**What's Next?**' section.

10

CONCLUSION (& A MESSAGE TO THE READER)

———◦———

We have come a long way since the 'World Turtle Theory' according to which the earth is flat and is resting on a tower of infinite turtles! (There are still some 'flat earthers' today, though). Today, we are living in a world full of modern ideas, like curved space, black holes, dark matter, multi-universe theory etc.

It is amazing to know how we, as an animal species, have become able to understand ourselves, our surroundings and our universe. At some time, we were monkey-like organisms, but our natural tendency of being curious has got us here. We have developed a lot since then. A recent example of this is the picture of the black hole. It is 55 million light years away from us, in another galaxy, but we still captured it. And, it is more interesting to know that the picture of the black hole is completely identical to the Astrophysicists' illustrations of a Black Hole (predicted using Einstein's General Relativity) which is a proof of the accuracy of our mathematics and our theories. It is also a very strong proof in the support of General Relativity.

We have developed in almost every field of science and technology, Physics being the basic and fundamental for all the development. And, the 20th century was a revolution in this development. Twentieth century revolutionized our ideas about space, time, matter, energy and the universe.

We have discussed in this book how a basic foundation was laid by Newton and Galileo for the modern ideas of space and time. These ideas were used by Albert Einstein to develop special relativity, which, in turn, is a foundation for the biggest two theories of modern physics — The General Theory of Relativity and Quantum Physics.

As we have already discussed that general relativity and quantum mechanics are incompatible with each other, there arises a need to obtain a unified theory of everything that would describe almost all the (observable) universe.

In chapter 8, we discussed how the universe started and how it will end (probably). But, where is God in this timeline and what is God's role in creating the universe?

If we think scientifically and with logic, there is no God. The universe is governed by the laws of physics which were defined at the time of the Big Bang. There are three simple reasons of why god doesn't exist:

1. No evidence

2. Illogical

3. We don't need a creator or god

There would be no time for that so-called creator to create the universe as space and time themselves started at the Big Bang [Now you would be thinking that I am just mentioning Hawking's words! But, really I myself realized that there is no god before I even knew Hawking's views on religion and god]. Really, ask yourself, do you need to believe in a god? Do you really need to believe that god does everything, not the laws of physics? Would you believe in god if no one ever told you about its existence? Well, there's a reason behind everything and we have to understand this thing! I am not saying or 'forcing' anyone to change his or her views on religion or god. At least, according to me, there is no god (and that's a fact!).

As we discover the laws of nature more and more, the need to believe in god decreases.

For example, many people believed (and some still believe!) that the solar and lunar eclipses take place when a demon (like a white wolf!) ate the 'sun god' or 'moon god'. So, when an eclipse took place, the people would start making as much noise as possible to 'scare' the demon. Some people also believed that to end the 'anger' of the sun god, a sacrifice of a person is needed!

But, as more and more physical laws were uncovered, everything became more clear. Everything has a reason. And, when we will obtain a unified theory of QM and GR, also known as 'Theory of Quantum Gravity' or 'Theory of Everything', the idea of God would seem even more ridiculous, because then we would know how the universe actually started, that is, what actually happened at the instant of Big Bang.

If scientists would believe that god does everything, we won't be living this life that we are living today.

It is a very interesting fact that we, like other objects in the universe, are just a group of fundamental particles, but we understand the universe. We are the only group of these fundamental particles who know that they are a group of fundamental particles!

But then we have all types of people. It is really very disappointing to see that on one hand people are doing so much hard work to find out the mysteries of the universe etc. and giving some real contribution to the world, and on the other hand we still have people who believe in religion, god and spirituality. We still have people who believe in a flat earth and we still have people who believe in astrology. (One thing many people don't know is that astrology and astronomy are two completely different things. Astronomy studies celestial objects and celestial phenomena using science. And, astrology is how stars and planets affect our life and affect our future. Obviously, any star or planet can't affect our life until it eventually collides with the earth and destroys everything!) Believing in all these things does nothing but slows down the development of the human race.

Another thing which is really dumb and slows down the development of our race is patriotism. It is a very narrow minded mentality. It is more like a political ideology than a scientific thing. Really, why would you be proud of being born in a particular piece of land? (The world is just divided into various countries to ease down the work of administration. A single person can't handle the whole world, that is the only reason that we have divisions). In this way, you can break it down on an even smaller level, like "I am proud to be born in this state, this city and I am proud to be born in this hospital!" You should be proud of something you achieve on your own in your life, not something that happens by an accident of birth. Being an American, British, Russian or Indian isn't an achievement; it's just a genetic accident. You wouldn't be proud to be born with 5 kilograms of weight at the time of birth, then why actually you would be proud to be born in America or India or anything? If you think logically and from a scientific point of view, all this patriotism and stuff is just meaningless, and it does nothing but just divide people and spreads hate among people.

Really, "We live in a society exquisitely dependent on science and technology, in which hardly anyone knows anything about science and technology." (— Carl Sagan).

I will just end this here by saying that people have to understand that there is nothing beyond science and logic. Everything has a reason.

Finally, I will end (the last chapter) this book with a quote of American Astronomer, Carl Sagan —

"The Cosmos is within us. We are made of Star-Stuff. We are a way for the universe to know itself."

WHAT'S NEXT?

So, now you've read all the stuff in this book and maybe you want to know more about all these things and really understand Physics.

I will just straightly describe to you the correct way of learning Physics (by yourself) from the basics to an advanced level, so you will have some technical knowledge of the subject.

This book may have given you a 'feel' of how the subject looks like and you may be introduced to the subject now.

So, now a thing you have to do is to get these textbooks - **"Feynman Lectures of Physics"** (3 volumes) by *Richard Feynman*, **"Fundamentals of Physics"** (2 volumes) by *Ramamurti Shankar* and **"Concepts of Physics"** (2 volumes) by *H.C. Verma*. Another thing you have to know before doing these books is "introductory calculus", a branch of mathematics which is often used in Physics.

These three books are technical books which will give you the real knowledge of Physics, which is important to understand the subject.

This is how you've to use these books:

1. Read a chapter from "Fundamentals of Physics".

2. Then read the same chapter from "Feynman Lectures of Physics".

3. Do the problems/questions/exercises of the chapter from "Fundamentals of Physics".

4. Do the problems/questions/exercises of the same chapter from "Concepts of Physics".

That is how you've to do every chapter.

After doing all these books, you will build a strong foundation in Physics. After that, you can get a (technical) textbook on any "specialized subject" you like, for example, if you want to learn general relativity and cosmology to an advanced level, get **"Introduction to General Relativity, Black Holes and Cosmology"** by *Yvonne Choquet-Bruhat* and so on.

ACKNOWLEDGEMENTS

I would like to thank all my family members and friends who supported and encouraged me while writing this book. I would also like to thank my parents for making this book better. I thank my friends and co-bloggers at Maddyz Physics — Siddavatam Sathvik, Sanchay Upadhyay, Nakul Kaushik and Sameer Pratap Singh for their consistent support, help and encouragement.

Writing this book wouldn't be possible for me without my friends and family.

This book is sponsored by **Maddyz Physics** and **Jalayoga India.**

SPONSORS

Maddyz Physics

Maddyz Physics is a science website founded by Madhur Sorout (me!) on January 15, 2018.

This project has been started "to make the world understand that there's a reason behind everything"!

The blog keeps the users updated with the latest science news and also features descriptive articles from a range of topics of Physics, Astronomy, Cosmology, Mathematics, Technology, Engineering and Chemistry for general readers (Laymen). It also provides daily science stuff to its audience on its social media handles.

It has gained an audience of **10,000+** (on social media) since September 2019.

Official Website – https://maddyzphysics.com

Instagram Page - @maddyzphysics - https://www.instagram.com/maddyzphysics/

Facebook Page - https://www.facebook.com/maddyzphysics/

Jalayoga India

A partner of Maddyz Physics, Jalayoga India is a YouTube channel which officially airs 'Jala Yog Divas'. Jala Yog Divas is organized by Jala Yogi **Shree Partap Singh Dagar** every year since 2015.

Official Link – https://bit.ly/jalayoga-india

REFERENCES AND SUGGESTED READINGS

- Chen, Susan. "The Basics of Special Relativity." Passion For STEM. May 16, 2017. https://passionforstem.wordpress.com/2017/01/02/the-basics-of- special-relativity

- Chen, Susan. "The De Broglie Hypothesis – Physics Research Project#1." Passion For STEM. August 14, 2018. https://passionforstem.wordpress.com/2018/05/31/the-de-broglie-hypothesis- physics-research-project-1/

- Feynman, Richard; Sands, Matthew; Leighton, Robert. The Feynman Lectures on Physics, Vol. 1: Mainly Mechanics, Radiation, and Heat. Dorling Kindersley (India), 2013

- Shankar, Ramamurti. Fundamentals of Physics. Yale University Press, 2019.

- S N Dhawan, S C Kheterpal, and P N Kapil. Pradeep's New Course Chemistry Class XI Volume 1 2019–20. Jalandhar: Pradeep Publications, 2019.

- Verma, H.C. Concepts of Physics. Part 1. New Delhi: Bharati Bhawan, 1999.

- Verma, H.C. Concepts of Physics. Part 2. New Delhi: Bharati Bhawan, 1999.

- Hawking, S W. A Brief History of Time : From the Big Bang to Black Holes. London: Bantam Books, 2016.

- Hawking, Stephen. Brief Answers to the Big Questions : Stephen Hawking., 2018.

- Hawking, Stephen W, and Leonard Mlodinow. A Briefer History of Time. New York: Bantam Dell, 2008.

- Kaul, Ish. The Physics of Fanfiction. Notion Press, 2017.

- Madhur Sorout. "Classical Physics in Brief." Maddyz Physics. Maddyz Physics, March 27, 2018. https://maddyzphysics.com/2018/03/classical-physics-in-brief/

- Madhur Sorout. "The Basics of Quantum Physics." Maddyz Physics. Maddyz Physics,

- April 21, 2018. https://maddyzphysics.com/2018/04/the-uncertainty-principle/

- Madhur Sorout. "Theory of Relativity in Brief." Maddyz Physics. Maddyz Physics, April 7,

2018. https://maddyzphysics.com/2018/04/theory-of-relativity-in-brief/

- Muller, Richard, and Shaun Maguire. "Now, and the Flow of Time." arXiv:1606.07975v1, 2016. https://arxiv.org/pdf/1606.07975

- Neil Degrasse Tyson. Astrophysics for People in a Hurry. New York Norton, 2017.

ABOUT THE AUTHOR

Madhur Sorout is a science blogger and some sort of a science communicator as well. He has founded the science website – Maddyz Physics (maddyzphysics. com). He has also been a physics and astrophysics editor for the Young Scientists Journal.

Other than physics, astronomy and mathematics, he takes interests in a number of other things mainly music, anime, fiction, cricket and the list goes on.

Social Media Handles of the Author:

- Instagram-@madhursorout&@maddyzphysics-https://www.instagram.com/madhursorout/ https://www.instagram.com/maddyzphysics/

- Facebook Page - https://www.facebook.com/ madhursorout

Printed in Great Britain
by Amazon

15359259R00130